Introduction to Matroids

Brian A. Kolo, Ph.D. J.D.

All questions and comments concerning this publication should be directed to publisher@weatherfordpress.com.

ISBN-13: 978-1-61580-010-0
ISBN-10: 1-61580-010-7

This book is dedicated to my father who patiently taught me everything about math, physics, and astronomy over countless breakfasts and starry evenings.

Special thanks to Saraa Elhagehammoud for the cover and review of the contents.

Table of Contents

Chapter 1: Preliminary Concepts

In this section we introduce some basic mathematical concepts that are used periodically in our study of matroids. The theory of sets and functions occur frequently, and the basic concepts are provided here both as an introduction to the material as well as a reference.

1.1 Sets

Sets are used extensively in this book and are fundamental to the definition of a matroid. This chapter introduces some basic set theory and set concepts that are used later.

1.1.1 DEFINITION: SET

A set is a collection of distinct objects.

1.1.2 DEFINITION: ELEMENT

Let S be a set. The objects that make up S are called elements or members.

1.1.3 EXAMPLE

If S is the set of colors in a rainbow, what are the elements of S?

The prismatic colors are: red, orange, yellow, green, blue, indigo, violet. These are the elements of the set S. S is designated as:

$$S = \{red, orange, yellow, green, blue, indigo, violet\}.$$

The items inside the braces are elements of the set, and the elements are separated by a comma. The braces indicate that the list is a set.

1.1.4 EXAMPLE

Find the set of positive integers less than five.

$$S = \{1,2,3,4\}$$

1.1.5   EXAMPLE

Determine which of the following is a set:

$$S_1 = \{0,1,2\}$$

$$S_2 = \{0,1,2,2\}$$

$$S_3 = \{orange, 0, 3.14, cat\}$$

$$S_4 = \{0, \{0\}\}$$

S_1 is a set of integers. This is a set because there is a list of elements, and the elements are distinct.

S_2 is not a set. There is a list of elements, but there are two elements with the value 2. Because all of the elements are not distinct, S_2 is not a set.

S_3 is a set. There is a list of elements, and the elements are all distinct. This is a set even though the elements are of different types.

S_4 is a set. The first element of S_4 has the value 0. The second element of S_4 is the set $\{0\}$. The value 0 is distinct from a set containing 0.

1.1.6   DEFINITION: EMPTY SET

Let S be a set. If S has no elements, S is called the empty set. The empty set is designated by $\emptyset$.

1.1.7   EXAMPLE

Find the set S of all real numbers x such that $x^2 = -1$.

There are no real numbers that satisfy $x^2 = -1$. Thus, the set S is empty, or $S = \emptyset$.

1.1.8   EXAMPLE

Find the set S of all real numbers x such that $x = x$.

All real numbers are equal to themselves. We can describe this with a set by:

$$S = \{x \mid x \in \mathbb{R}\}.$$

This is read: S = the set ({ }) of x's such that (|) x is contained ($\in$) in the real numbers ($\mathbb{R}$).

1.1.9 DEFINITION: CARDINALITY

> Let S be a finite set. The cardinality of S is the number of elements in S. Cardinality is designated by $|S|$.

1.1.10 EXAMPLE

Find the cardinality of the sets in Examples 1.1.3 and 1.1.4.

In example 1.1.3, the set of colors has seven elements, so $|S| = 7$.

For example 1.1.4, the set contains four elements, so $|S| = 4$.

1.1.11 EXAMPLE

Find the cardinality of the sets in example 1.1.5.

Only three of the items in Example 1.1.5 are sets. The cardinalities of the sets are:

$$|S_1| = 3$$

$$|S_3| = 4$$

$$|S_4| = 2$$

1.1.12 EXAMPLE

Find the cardinality of $S = \emptyset$.

There are zero elements in S, so $|S| = 0$.

1.1.13 EXAMPLE

Find the cardinality of $S = \{\emptyset\}$.

There is one element in S, so $|S| = 1$. Compare this with the previous example. The cardinality of the empty set is zero. However, the cardinality of a set that has one member which is the empty set, is one.

This highlights the difference between the empty set as a set versus the empty set as an element of a set.

1.1.14 DEFINITION: CARDINALITY OF THE INTEGERS

> Let $\mathbb{Z}$ be the set of integers. There are an infinite number of elements in $\mathbb{Z}$ and the cardinality of this set is $\aleph_0$ (aleph-null).

Any set that can be placed in one-to-one correspondence with the integers has cardinality $\aleph_0$.

1.1.15 DEFINITION: CARDINALITY OF THE CONTINUUM

> Let $\mathbb{R}$ be the set of real numbers. The cardinality of this set is $\mathbb{c} = 2^{\aleph_0} = \beth_1$ (beth-one).

1.1.16 PROBLEMS

1. Find the cardinality of the following sets:
   a. $S = \{1,2,5\}$
   b. $S = \{-1,0,1,2,3,4\}$
   c. $S = \{all\ x \mid x^2 = 2\}$
2. What is the set of vowels? What is the cardinality of this set?

1.2   Subsets

Subsets are sets that are contained in another set. Subsets are very important in matroid theory as the fundamental definitions of a matroid involve the relationships between subsets.

1.2.1 DEFINITION: SUBSET

> Let S be a set. If S' is a subset of S, then every element of S' must also be an element of S. If S' is a subset of S, we write $S' \subseteq S$.

From the definition of a set, every set is a subset of itself: $S \subseteq S$.

1.2.2 EXAMPLE

Find three subsets of the set $S = \{ginger, spice\}$.

The set $S_1 = \{ginger, spice\}$ is a subset of S. Every element in S_1 is also an element of S.

The set $S_2 = \{ginger\}$ is a subset of S. Every element in S_2 is also an element of S.

The set $S_3 = \emptyset$ is a subset of S. There are no elements in S_3. Thus, every element in S_3 is also in S.

1.2.3   DEFINITION: PROPER SUBSET

> Let S be a set. S' is a proper subset if $S' \subseteq S$, and there exists at least one element of S that is not contained in S'. If S' is a proper subset of S, we write $S' \subset S$.

1.2.4   EXAMPLE

Find all subsets of the set $S = \{a, b, c\}$. Which of these subsets is proper?

We organize the subsets by cardinality.

$|S'| = 3$

There is one subset of cardinality three: $S' = \{a, b, c\}$. This is not a proper subset because there is no element in S that is not in S'.

$|S'| = 2$

There are three subsets of cardinality two:

$$\{a, b\} \quad \{a, c\} \quad \{b, c\}$$

All of these sets are proper because in each case there is some element in S that is not in S'.

$|S'| = 1$

There are three subsets of cardinality one:

$$\{a\} \quad \{b\} \quad \{c\}$$

All of these sets are proper because in each case, there is some element in S that is not in S', and $S' \neq \emptyset$.

$|S'| = 0$

There is one subset of cardinality zero: $S' = \emptyset$. This is a proper subset.

1.2.5   DEFINITION: UNION

> Let S and T be sets. The union U of S and T is the set containing all elements of S and all elements of T. The union is designated $U = S \cup T$.

The union is a method to combine two sets together.

1.2.6   EXAMPLE

Find the union of the sets from Examples 1.2.2 and 1.2.4.

$$U = \{ginger, spice\} \cup \{a, b, c\} = \{ginger, spice, a, b, c\}$$

1.2.7   EXAMPLE

Find the union of the sets

$$S = \{a, b, c, d\} \quad T = \{a, b, e, f\}.$$

To form the union we make a set containing every element of S and every element of T. Elements that are common to both S and T are only in the union once (because the union is a set). In this example,

$$U = S \cup T = \{a, b, c, d, e, f\}.$$

1.2.8   PROPOSITION

If S and T are sets with $|S| = s$ and $|T| = t$, then the cardinality of the union of S and T satisfies

$|S \cup T| \leq s + t.$

Proof. If S and T have no elements in common, then the union of S and T will have s elements from S and t elements from T. In this situation, the cardinality of the union is $s + t$. However, any elements that are in both S

and T are present in the union only once. In this case the cardinality is less than $s + t$.∎

1.2.9 DEFINITION: INTERSECTION

> Let S and T be sets. The intersection I of S and T is the set containing all elements that are present in both S and T. The intersection is designated $I = S \cap T$.

The intersection is a method to find the common elements between two sets. Two sets that have no elements in common are called disjoint.

1.2.10 EXAMPLE

Find the intersection of the sets from Examples 1.2.2 and 1.2.4.

$$I = \{ginger, spice\} \cap \{a, b, c\} = \emptyset$$

1.2.11 EXAMPLE

Find the intersection of the sets

$$S = \{a, b, c, d\} \quad T = \{a, b, e, f\}.$$

There are two elements in common between these sets:

$$I = S \cap T = \{a, b\}.$$

1.2.12 PROPOSITION

If S and T are sets then $I = S \cap T = T \cap S$.

Proof. The intersection $S \cap T$ is the set of elements common to both S and T. If we write the intersection as $T \cap S$, we arrive at the same set of elements, the elements common to both T and S.∎

1.2.13 DEFINITION: SET DIFFERENCE

> Let S and T be sets. The difference $D = S - T$ is the set containing all elements of S that are not in T.

1.2.14 EXAMPLE

Find the difference of the sets from examples 1.2.2 and 1.2.4.

There are no common elements in these sets, so the difference is just the original set.

$$D_1 = \{ginger, spice\} - \{a, b, c\} = \{ginger, spice\}$$

$$D_2 = \{a, b, c\} - \{ginger, spice\} = \{a, b, c\}$$

1.2.15 EXAMPLE

Find the difference of the sets

$$S = \{a, b, c, d\} \quad T = \{a, b, e, f\}.$$

There are some common elements between these sets, so the difference is not simply the original set:

$$D_1 = S - T = \{c, d\}$$

$$D_2 = T - S = \{e, f\}$$

1.2.16 PROPOSITION

If S and T are sets then $S - T$ may be different from $T - S$.

Proof. The previous examples provide explicit examples where $S - T$ is different from $T - S$. ∎

1.2.17 EXAMPLE

Figure 1 shows a Venn diagram for the intersection, union, and difference of two sets. Each circle is meant to represent a set. The union of the sets is everything contained in both circles. The intersection is the common, crossed-hatched area. The difference $A - B$ is the left area of A but removing any part common to both A and B. Similarly, difference $B - A$ is the right area of B but removing any part common to both A and B.

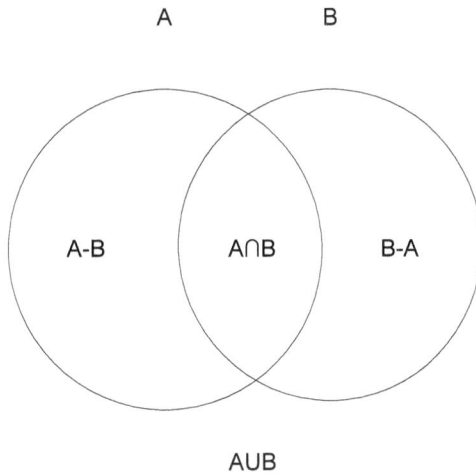

Figure 1: Venn diagram of set intersection and difference.

1.2.18 PROPOSITION

Let A and B be sets. Let $I = A \cap B$ and $D = A - B$. Then $I \cap D = \emptyset$, and $A = I \cup D$.

Proof. Examining the Venn diagram from Figure 1, it appears that the sets $A - B$ and $A \cap B$ do not have any elements in common. This is in fact the case because none of elements in the set $A - B$ are in B, while all of the elements in the set $A \cap B$ are in B.

Furthermore, we can classify the elements of A into two categories: 1) those elements of A that are not in B, and 2) those elements of A that are in B. The set A is the union of these two sets:

$$A = (A - B) \cup (A \cap B). \blacksquare$$

1.2.19 PROPERTIES

We state without proof some important properties of sets. These properties are used throughout the remainder of this book. Let A, B, and C be sets that are all subsets of S.

Union	Intersection	Name
$A \cup (A \cup C) = (A \cup B) \cup C$	$A \cap (A \cap C) = (A \cap B) \cap C$	**SP-1** Associative
$A \cup B = B \cup A$	$A \cap B = B \cap A$	**SP-2** Commutative
$A \cup (B \cup C) = (A \cup B) \cup (A \cup C)$	$A \cup (B \cup C) = (A \cup B) \cup (A \cup C)$	**SP-3** Distributive
$A \cup \emptyset = A$	$A \cap S = A$	**SP-4** Identity
$A \cup (S - A) = S$	$A \cap (S - A) = \emptyset$	**SP-5** Complement
$A \cup A = A$	$A \cap A = A$	**SP-6** Idempotent
$A \cup (A \cap B) = A$	$A \cap (A \cup B) = A$	**SP-7** Absorption

1.2.20 DEFINITION: SYMMETRIC DIFFERENCE

Let S and T be sets. The symmetric difference is defined as

$$S \Delta T = (S \cup T) - (S \cap T)$$

1.2.21 EXAMPLE

Find the symmetric difference of the sets $S = \{a, b, c, d\}$ and $T = \{b, c, e, f\}$.

$$S \Delta T = (S \cup T) - (S \cap T)$$

$$= \{a, b, c, d\} \cup \{b, c, e, f\} - \{a, b, c, d\} \cap \{b, c, e, f\}$$

$$= \{a, b, c, d, e, f\} - \{b, c\}$$

$$= \{a, d, e, f\}$$

1.2.22 PROBLEMS

1. Let $S = \{a, b\}$.
   a. Find every subset of S.
   b. What are the proper subsets of S?
2. Let $S = \{-1, 0, 1\}$ and $T = \{0, 1, 2\}$. Find:
   a. $S \cup T$
   b. $S \cap T$
   c. $S - T$
   d. $T - S$
   e. $S \Delta T$

3. Let $S = \{a, b, c, d\}$ and $T = \{b, c\}$. Find:

   a. $S \cup T$
   b. $S \cap T$
   c. $S - T$
   d. $T - S$
   e. $S \Delta T$

1.3   Power Sets

Power sets are a method of generating a set of sets given an initial set. The power set of a set S is a set whose elements are all subsets of S. The power set is not restricted to proper subsets, so both S and $\emptyset$ are members of the power set.

1.3.1   DEFINITION: POWER SET

Let S be a set. The power set P of S is the set of all subsets of S.

1.3.2   EXAMPLE

Find the power set of the set $S = \{a, b, c\}$.

The subsets of S are: S, $\{a, b\}, \{a, c\}, \{b, c\}, \{a\}, \{b\}, \{c\}, \emptyset$. The power set of S is

$$P = \{\emptyset, \{a\}, \{b\}, \{c\}, \{a, b\}, \{a, c\}, \{b, c\}, \{a, b, c\}\}$$

In set notation, $\{a, b\} \in P$, but $\{a, b\} \subset S$. This means the set $\{a, b\}$ is actually a member of P. However, $\{a, b\}$ is not a member of S. Because both a and b are members of S, the set $\{a, b\}$ is a subset of S, but the set $\{a, b\}$ is not actually a member of S. The members of S are a, b and c.

1.3.3   THEOREM

Let S be a finite set and P be the power set of S. Then $|P| = 2^{|S|}$.

1.3.4   EXAMPLE

Find the power set of the set $S = \{a, \{b\}\}$.

The subsets of S are: S, $\{a, \{b\}\}, \{a\}, \{\{b\}\}, \emptyset$. The power set of S is

$$P = \left\{ \emptyset, \{a\}, \{\{b\}\}, \{a, \{b\}\} \right\}$$

In this case, the set S has elements which are sets themselves. The power set is formed just as before, but the power set formed also has elements which are themselves sets. Careful distinction should be made between a set element $a \in S$, and an element which is also a set $\{b\} \in S$.

In the last example, $a \in S$, but $a \notin P$. Similarly, $\{a\} \notin S$, but $\{a\} \in P$. It is important to distinguish when a set is a *member* of set ($\in$), and when a set is a *subset* of a set ($\subset$). This distinction is often overlooked and can be confusing when set notation is not strictly followed.

1.3.5   EXAMPLE

Find the power set of the set $S = \{\emptyset\}$.

First, note that $S \neq \emptyset$. Rather, S is a set containing $\emptyset$ as a member. There are only two subsets for S: $\{\emptyset\}$ and $\emptyset$. Again, there is a difference between the set $\emptyset$, and a set $\{\emptyset\}$ that has $\emptyset$ as an element. The power set of S is

$$P = \left\{ \emptyset, \{\emptyset\} \right\}$$

1.3.6   THEOREM

Let S and T be finite sets with $S \cap T = \emptyset$. Let P be the power set of $S \cup T$. Then $|P| = 2^{|S|+|T|}$.

Proof. From theorem 1.3.3, the dimension of a power set is $|P| = 2^{|S|}$. Since $S \cap T = \emptyset$, there are no elements in common between the sets S and T. In this case we have

$$|S \cup T| = |S| + |T|.$$

Then the dimension of the power set of $S \cup T$ is

$$|P| = 2^{|S \cup T|} = 2^{|S|+|T|}. \blacksquare$$

1.3.7   PROBLEMS

1.   Prove Theorem 1.3.3.

1.4   Maximal and Minimal Members of a Set

The concepts of maximal and minimal members of a set are important for defining certain aspects of matroids in later sections. This section will provide the fundamental definitions for maximal and minimal sets along with several illustrative examples.

1.4.1   DEFINITION: MAXIMAL MEMBERS OF A SET

> Let S be a set. The maximal members of A are the elements of A that are not proper subsets of any member of A. The set of maximal members of A is designated $[A]$.

1.4.2   EXAMPLE

Find the maximal members of the set $S = \{\{a\}, \{b\}, \{a, b\}\}$.

Examine each element of S :

{a}

The element $\{a\}$ is a proper subset of the element $\{a, b\}$, so $\{a\}$ is not a maximal member of S.

{b}

Similarly, the element $\{b\}$ is also proper subset of the element $\{a, b\}$, so $\{b\}$ is not a maximal member of S.

{a, b}

The element $\{a, b\}$ is not a proper subset of $\{a\}$, $\{b\}$, or $\{a, b\}$. Thus, $\{a, b\}$ is a maximal member of S.

Thus, $[S] = \{\{a, b\}\}$.

1.4.3   PROPOSITION

Let S be a set and $P = 2^E$ be the power set of S. Then P has a unique element that is a maximal member.

Proof. The power set of S is the set of all subsets of S. The set S is a member of P, and all other members of P are subsets of S. Thus, S is the only maximal member of P. ∎

1.4.4   EXAMPLE

Consider the set $S = \{a, b, c\}$. The power set P of S is

$$P = \{\emptyset, \{a\}, \{b\}, \{c\}, \{a, b\}, \{a, c\}, \{b, c\}, \{a, b, c\}\}.$$

Examining each element of P:

$$\emptyset \subset \{a, b, c\} \quad \{a\} \subset \{a, b, c\} \quad \{b\} \subset \{a, b, c\} \quad \{c\} \subset \{a, b, c\}$$

$$\{a, b\} \subset \{a, b, c\} \quad\quad \{a, c\} \subset \{a, b, c\} \quad\quad \{b, c\} \subset \{a, b, c\}$$

Finally, $\{a, b, c\} \not\subset \{a, b, c\}$, so $\{a, b, c\}$ is a maximal element of P.

The member $\{a, b, c\}$ is the single unique maximal member of P. This member is equivalent to the original set S. In accordance with Proposition 1.4.3, the maximal member of the power set is the original generating set.

1.4.5   DEFINITION: MINIMAL MEMBERS OF A SET

Let S be a set. The minimal members of A are the elements of A that do not contain any member of A proper subset. The set of minimal members of A is designated $\lfloor A \rfloor$.

1.4.6   EXAMPLE

Find the minimal members of the set $S = \{\{a\}, \{b\}, \{a, b\}\}$.

Examine each element of S:

$$\{a\}$$

No element of S is a proper subset of $\{a\}$, so $\{a\}$ must be a minimal member of S.

{**b**}

Similarly, no element of S is a proper subset of $\{b\}$, so $\{b\}$ is a minimal member of S.

{**a**, **b**}

The element $\{a, b\}$ has both $\{a\}$ and $\{b\}$ as proper subsets. In addition, both $\{a\}$ and $\{b\}$ are members of S, so $\{a, b\}$ is not a minimal member of S.

Thus, $\lfloor A \rfloor = \{\{a\}, \{b\}\}$.

1.4.7 EXAMPLE

Find the minimal elements of $S = \{\{a\}, \{a, b\}, \{b, c\}, \{b, c, d\}\}$.

{**a**}

No element of S is a proper subset of $\{a\}$, so $\{a\}$ must be a minimal member of S.

{**a**, **b**}

The element $\{a, b\}$ has both $\{a\}$ and $\{b\}$ as proper subsets. Since $\{a\}$ is a subset of $\{a, b\}$ and $\{a\}$ is a member of S, $\{a, b\}$ is not a minimal member of S.

{**b**, **c**}

The element $\{b, c\}$ has both $\{b\}$ and $\{c\}$ as proper subsets. However, neither $\{b\}$ nor $\{c\}$ are elements of S. $\{b, c\}$ is a minimal element of S.

{**b**, **c**, **d**}

Element $\{b, c, d\}$ has $\{b, c\}$ as a proper subset. But $\{b, c\}$ is also a member of S, so $\{b, c, d\}$ is not a minimal element of S.

1.4.8 PROBLEMS

1. Find the maximal and minimal members of the set
   $$S = \{\{a\}, \{b\}, \{c\}, \{a, b\}, \{b, c\}\}$$

1.5   Functions

Functions on sets play an important role throughout mathematics. This section presents some basic theory of functions.

1.5.1   DEFINITION: FUNCTION

> Let A and B be sets. A function is a mapping between every element of A and the elements of B. We denote the function f by
>
> $$f : A \rightarrow B$$

1.5.2   EXAMPLE

Let A and B be the sets

$$A = \{a, b, c\} \quad B = \{1, 2, 3\}.$$

We can specify f as a set of ordered pairs indicating that an element of A is mapped to an element of B:

$$(a, 1) \quad (b, 2) \quad (c, 3)$$

Another way to express the function is with ( )'s:

$$f(a) = 1 \quad f(b) = 2 \quad f(c) = 3$$

1.5.3   DEFINITION: INJECTION

> Let A and B be sets. A function $f : A \rightarrow B$ is an injection if each element of A is mapped to a unique element of B.

Injective functions are also called 'one-to-one'. Figure 2 shows an injective function. Note that every element of A is used, but we do not need to use every element of B.

Injective functions can be reversed since each element of A is mapped to a unique element of B. We can imagine writing down every ordered

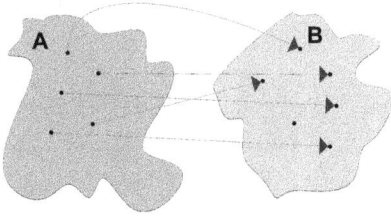

Figure 2: Injection between two sets.   Figure 3: Surjection between two sets.

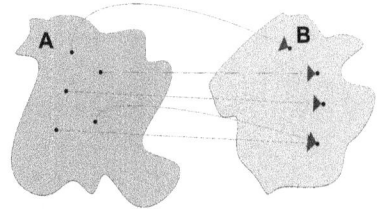

pair (a_i, b_i), then reversing these to get (b_i, a_i). The new order pairs reverses f in the sense that we travel against the arrows in Figure 2.

However, the new set of ordered pairs is not necessarily a function. In order for this to be a function, every element of B would need to be traced back to an element of A. But the injection is not required to use every element of B, that is, every element of B may not be mapped to an element of A.

1.5.4   DEFINITION: SURJECTION

> Let A and B be sets. A function $f: A \rightarrow B$ is a surjection if each element of B corresponds to at least one element of A.

Surjective functions are also called 'onto'. In a surjection, every element of B is used. Every element of A is mapped to some element of B, and there are no elements of B left out. Figure 3 shows a surjective function.

It is not necessarily possible to reverse a surjective function. As Figure 3 indicates, we can have more than one value of A mapping to a single value of B. In this case, give an element B, we cannot identify a unique value of A. We have multiple possible values of A, and an inverse function requires a unique value.

1.5.5   EXAMPLE

Is the function from Example 1.5.2 injective and/or surjective?

From the definition of the function, we see that every element of A is mapped to some element of B so f is in fact a function. We can see that every element of A is mapped to a unique element of B because no value of B is used more than once. This makes f an injection. Furthermore, every element of B is used as well, so f is a surjection.

1.5.6   DEFINITION: BIJECTION

> Let A and B be sets. A function $f : A \rightarrow B$ is a bijection if f is both a surjection and an injection.

Although an injection uniquely maps every element of A, we cannot reverse an injection because we didn't use every element of B. In addition, a surjection uses every element of B, but we cannot reverse a surjection because it does not uniquely map every element of A.

However, if we put these two together, we get all the properties we need to guarantee we can create in inverse function. An injection guarantees that every element of B is used, and a surjection guarantees that every pair (a_i, b_i) is unique.

1.5.7   EXAMPLE

Let A and B be the sets

$$A = \{a, b, c\} \quad B = \{1,2,3\}.$$

Let $f : A \rightarrow B$ be a function defined as:

$$(a, 1) \quad (b, 2) \quad (c, 1)$$

Describe f in terms of injective, surjective, and bijective.

f is not injective because f maps both a and c to 1.

f is not surjective because f does not use element 3 of B.

f is not bijective because f is neither surjective not injective.

1.5.8   EXAMPLE

Let A and B be the sets

$$A = \{a, b, c, d\} \quad B = \{1,2,3\}.$$

Let $f : A \rightarrow B$ be a function defined as:

$$(a,1) \quad (b,2) \quad (c,1) \quad (d,3)$$

Describe f in terms of injective, surjective, and bijective.

f is not injective because f maps both a and c to 1.

f is surjective because f maps every element of B.

f is not bijective because f is not injective.

1.5.9   EXAMPLE

Let A and B be the sets

$$A = \{a, b, c\} \quad B = \{1,2,3,4\}.$$

Let $f: A \rightarrow B$ be a function defined as:

$$(a,1) \quad (b,2) \quad (c,4)$$

Describe f in terms of injective, surjective, and bijective.

f is injective because f maps every element of A to a unique element of B.

f is not surjective because f does not use element 3 of B.

f is not bijective because f is not surjective.

1.5.10   EXAMPLE

Let A and B be the sets

$$A = \{a, b, c, d\} \quad B = \{1,2,3,4\}.$$

Let $f: A \rightarrow B$ be a function defined as:

$$(a,2) \quad (b,4) \quad (c,3) \quad (d,1)$$

Describe f in terms of injective, surjective, and bijective.

f is injective because f maps every element of A to a unique element of B.

f is surjective because f maps every element of B.

f is bijective because f is both injective and surjective.

1.5.11 THEOREM

Let A and B be finite sets with $|A| = |B|$. Let $f: A \rightarrow B$ be an injective function. Then f is bijective.

Proof. If f is injective, then every element of A maps to a unique element of B. However, since there are the same number of elements in both A and B, then we must use every element in B. But if we use every element in B, then f is surjective. Moreover, since f is both injective and surjective, f is bijective. ■

1.5.12 THEOREM

Let A and B be finite sets with $|A| = |B|$. Let $f: A \rightarrow B$ be a surjective function. Then f is bijective.

Proof. If f is surjective, then every element of B is used in f. But just like the above theorem, since there are the same number of elements in both A and B, and since every element of A is used, then each element of A must map to a unique element of B. This means that A is also injective. Again, since f is both injective and surjective, f is bijective. ■

1.5.13 PROBLEMS

1. Let $A = \{a, b, c, d\}$ and $B = \{1,2,3,4\}$. Let $f: A \rightarrow B$ be the map defined by the ordered pairs
$$(a, 2) \quad (b, 1) \quad (c, 3) \quad (d, 3)$$
Characterize this map as an injection, surjection, or bijection.

2. Let $A = \{a, b, c, d\}$ and $B = \{1,2,3,4\}$. Let $f: A \rightarrow B$ be the map defined by the ordered pairs
$$(a, 1) \quad (b, 3) \quad (c, 3) \quad (c, 4)$$
Characterize this map as an injection, surjection, or bijection.

3. Let $A = \{a, b, c, d\}$ and $B = \{1,2,3,4\}$. Let $f: A \rightarrow B$ be the map defined by the ordered pairs
$$(a, 3) \quad (b, 4) \quad (c, 2) \quad (c, 1)$$
Characterize this map as an injection, surjection, or bijection.

1.6   Homomorphisms

Homomorphisms are structure preserving maps. A homomorphism is a function that preserves some special relationship in the set. There are three main homomorphisms: monomorphisms, epimorphisms, and isomorphisms. These are structure preserving maps that are injective, surjective, and bijective respectively.

1.6.1.1 DEFINITION: HOMOMORPHISM

> Let A and B be sets. Let $f: A \rightarrow B$. f is a homomorphism if f preserves some structure of A.

A homomorphism preserves some set structure over the map. This structure preserving property allows the important properties of one set to transfer to another set. It is not necessary that every property transfer in this way, only the properties of interest.

1.6.2   EXAMPLE

Let A and B be the sets

$$A = \{a, b, c, d\} \quad B = \{e, f, g, h\}.$$

Let A and B be posets with the order relations

A:

$\leq$	a	b	c	d
a	1	1		1
b		1		1
c			1	
d				1

B:

$\leq$	e	f	g	h
e	1	1		1
f		1		1
g			1	
h				1

Let f be the map

$$a \rightarrow e \quad b \rightarrow f \quad c \rightarrow g \quad d \rightarrow h$$

f is a homomorphism of the order relation because after transforming the original order relation for A we have exactly the order relation for B.

≤	e	f	g	h
e	1	1		1
f		1		1
g			1	
h				1

The transformed order relation is exactly the same as the previous order relation, we have merely relabeled the set elements. In this sense, the function is a homomorphism because the ordering property is preserved.

However, note that we can spell the word 'bad' using set elements in set A. The transformed set B is not able to spell 'bad' using its set elements. The function f is not a homomorphism with respect to spelling.

1.6.3 DEFINITION: MONOMORPHISM

Let A and B be sets, and let $f: A \to B$ be a homomorphism. f is a monomorphism if f is an injection.

1.6.4 DEFINITION: EPIMORPHISM

Let A and B be sets, and let $f: A \to B$ be a homomorphism. f is an epimorphism if f is a surjection.

1.6.5 DEFINITION: ISOMORPHISM

Let A and B be sets, and let $f: A \to B$ be a homomorphism. f is an isomorphism if f is a bijection.

1.6.6 DEFINITION: ENDOMORPHISM

Let A be a set, and let $f: A \to A$. f a homomorphism. This special type of homomorphism is called an endomorphism.

An endomorphism is a special type of homomorphism that occurs when the function f maps the set A to itself rather than a different set.

1.6.7   DEFINITION: AUTOMORPHISM

> Let A be a set, and let $f: A \to A$. f an isomorphism. This special type of isomorphism is called an automorphism.

An automorphism is the particular case of an isomorphism where the isomorphism is mapping the set back to itself.

1.6.8   EXAMPLE

Reconsidering Example 1.6.2, the homomorphic map f is also a monomorphism because f is injective. Moreover, f is also an epimorphism because f is surjective. Furthermore, f is an isomporphism because f is a homomorphism that is a bijection.

However, f is neither an endomorphism nor an automorphism. f is a map between two different sets, A and B. f would need to be a map from A back to itself in order for f to be either an endomorphism or an automorphism.

1.6.9   EXAMPLE

Let A and B be the sets

$$A = \{a, b, c, d, e\} \quad B = \{a, b, c, d, e\} = A.$$

Let A and B be posets with the order relations

A:

$\leq$	a	b	c	d	e
a	1		1	1	1
b		1	1	1	1
c			1	1	1
d				1	
e					1

B:

$\leq$	b	c	d	e	a
b	1		1	1	1
c		1	1	1	1
d			1	1	1
e				1	
a					1

Let f be the map

$$a \to b \quad b \to c \quad c \to d \quad d \to e \quad e \to a$$

f is a homomorphism of the order relation because the transformed order relation satisfies the same order relation as the B:

≤	b	c	d	e	a
b	1		1	1	1
c		1	1	1	1
d			1	1	1
e				1	
a					1

Note that in the original set we have $a < c$ but in the new set $c < a$. We say that the transformed set preserves the order operation because the new set elements satisfy the new relation, not because they satisfy the old relation.

Specifically, if $a < c$, then $f(a) < f(c)$. We see that this is true because $f(a) = b$ and $f(c) = d$, and in the new relation, $b < d$. It is in this sense that the order relation is preserved.

In addition, f is both an injection and a surjection. So f is a monomorphism and an epimorphism. But f is also a bijection, so f is an isomorphism.

Moreover, f is an endomorphism because f is a homomorphism that maps A back to itself. Furthermore, f is an automorphism because f is both a homomorphism and an isomorphism.

1.7   Logic

We use some basic logic principles in later proofs. This section outlines some basic principles of logic statements and their relations.

1.7.1   DEFINITION: NEGATION

> The negation of a logical variable A is designated symbolically with an overbar $\bar{A}$.

1.7.2   DEFINITION: IMPLICATION

> An implication is a logic statement of the form if A then B. We designate this symbolically as
>
> $$A \rightarrow B$$

1.7.3   DEFINITION: INVERSE

The inverse of the statement $A \rightarrow B$ is the statement

$$\bar{A} \rightarrow \bar{B}$$

1.7.4   DEFINITION: CONVERSE

The converse of the statement $A \rightarrow B$ is the statement

$$B \rightarrow A$$

1.7.5   DEFINITION: CONTRAPOSITIVE

The contrapositive of the statement $A \rightarrow B$ is the statement

$$\bar{B} \rightarrow \bar{A}$$

1.7.6   PROPOSITION

If an implication is true, then the contrapositive is also true. The inverse and converse are not necessarily true. However, if either the converse or inverse is true, then the other is true also. ∎

1.7.7   EXAMPLE

To illustrate these relations, interpret A as 'fire' and B as 'hot'.

Implication:

$A \rightarrow B$ is read 'if fire, then hot'. Alternatively we might say 'fire is hot'. This implication is generally true.

Inverse:

$\bar{A} \rightarrow \bar{B}$ is read 'if not fire, then not hot'. Alternatively we might say 'if it's not fire, then it's not hot'. This is not generally true. Things like stoves, plasmas, and spicy food may be hot but none of these are fire.

Converse:

$B \rightarrow A$ is read 'if hot, then fire. Alternatively we might say 'if it's hot, then it must be fire'. This is not generally true. Similar to the Inverse above, stoves, plasmas, and spicy food may be hot but none of these are fire.

Contrapositive:

$\bar{B} \rightarrow \bar{A}$ is read 'if not hot, then not fire. Alternatively we might say 'if it's not hot, then it can't be fire'. This is generally true.

1.7.8   DEFINITION: AND / OR

> Logic often uses the concepts of AND and OR in crafting logic statements. These are represented with the symbols $\land$ for AND and $\lor$ for OR.

1.7.9   DEFINITION: DE MORGAN'S LAWS

> de Morgan's laws are binary implications between logical variables:
>
> **DM-1** $\overline{A \lor B} \Longleftrightarrow \bar{A} \land \bar{B}$
>
> **DM-2** $\overline{A \land B} \Longleftrightarrow \bar{A} \lor \bar{B}$

1.7.10   EXAMPLE

From the implication

$$A \land B \rightarrow C$$

find the inverse, converse, and contrapositive.

Inverse

From the definition of the inverse,

$$\overline{A \land B} \rightarrow \bar{C}.$$

Using de Morgan's law **DM-2**,

$$\bar{A} \lor \bar{B} \rightarrow \bar{C}.$$

Converse

From the definition of the converse,

$$C \to A \wedge B.$$

Contrapositive

From the definition of the contrapositive,

$$\bar{C} \to \overline{A \wedge B}.$$

Again, applying de Morgan's law **DM-2**,

$$\bar{C} \to \bar{A} \vee \bar{B}.$$

1.8   Transversals

A transversal may be thought of as set representing a particular set of choices. More formally, a transversal of a collection of sets is a set containing a subset from each set in the collection. Transversals prove useful in matching problems and will later form a class of matroids.

1.8.1   DEFINITION: TRANSVERSAL

Let S be a set, and A be a sequence of subsets where $A_i \subseteq S$ (not necessarily distinct). A transversal of S is a subset $T \subseteq S$ satisfying:

TV-1: $t_i \in A_i$ for each $t_i \in T$

TV-2: $|T| = |A|$

TV-3: $t_i \neq t_j$ for every $t_i, t_j \in T$ with $i \neq j$

A transversal is a sequence of choices from some set. At each step of the sequence, we may have a different set of options. We put all of the option choices together to form S.

A transversal is a specific set of these choices. However, when forming a transversal we are not allowed to pick the same option twice.

An example is color coordinating a wardrobe. Say we have shirts, pants, and shoes. We have shirts in the colors red, blue, or green. Pants are colored red, orange, black, or white. Shoes are white or gold. A transversal of our wardrobe is a selection of a shirt, pants, and shoes such that each is a different color.

1.8.2   EXAMPLE

Find a transversal of $S = \{a, b, c\}$ given $A = \{\{a, b\}, \{a, c\}, \{a, b\}\}$.

For convenience, designate $A = \{t_1, t_2, t_3\}$ where $t_1 = \{a, b\}$, $t_2 = \{a, c\}$, and $t_3 = \{a, b\}$.

There are only two sequences leading to a transversal for this system. We can specify these by trying every set of choices.

First, choose element a from t_1. Now we must choose c from t_2 because we have already used a and we are not allowed to use same choice twice. Finally, we must choose b from t_3. Since we have a set of choices where each choice is different, we have created a transversal $\{a, b, c\}$.

Repeating this process for all possibilities we find:

$$T: t_1 = a, t_2 = c, t_3 = b \rightarrow T = \{a, b, c\}$$

$$T: t_1 = b, t_2 = c, t_3 = a \rightarrow T = \{a, b, c\}$$

These is only one transversal: $\{a, b, c\}$. Note there are as many elements in the transversal as there are in A due to **T2**.

1.8.3   EXAMPLE

Find a transversal of the system

$$S = \{a, b, c, d\}$$

$$A = \{\{a, b, c\}, \{a\}, \{c, d\}, \{a, b, d\}\}.$$

Again, there are two sequences leading to transversals, and only one transversal:

$$T: t_1 = b, t_2 = a, t_3 = c, t_4 = d \rightarrow T = \{a, b, c, d\}$$

$$T: t_1 = c, t_2 = a, t_3 = d, t_4 = b \rightarrow T = \{a, b, c, d\}$$

1.8.4   EXAMPLE

Find a transversal of the system

$$S = \{a, b, c, d\}$$

$$A = \{\{a, b, c\}, \{a\}, \{c, d\}\}.$$

Here, there are three sequences leading to transversals:

$$T: t_1 = b, t_2 = a, t_3 = c \rightarrow T = \{a, b, c\}$$

$$T: t_1 = b, t_2 = a, t_3 = d \rightarrow T = \{a, b, d\}$$

$$T: t_1 = c, t_2 = a, t_3 = d \rightarrow T = \{a, c, d\}$$

These are the three transversals for this system.

1.8.5   DEFINITION: PARTIAL TRANSVERSAL

Let S be a set, and A be a sequence of subsets where $A_i \subseteq S$ (not necessarily distinct). Let $B \subseteq A$ be a subsequence of A. A partial transversal of S is a subset $P \subseteq S$ satisfying:

PT-1: $p_i \in B_i$ for each $p_i \in P$

PT-2: $|P| = |B|$

PT-3: $p_i \neq p_j$ for every $p_i, p_j \in P$ with $i \neq j$

PT-4: $|B| < |A|$

A partial transversal is similar to a transversal, only the subset in the partial transversal set is smaller than the sequence A.

1.8.6   EXAMPLE

Find a partial transversal of the system

$$S = \{a, b, c\}$$

$$A = \{\{a, b\}, \{a\}, \{c, d\}\}.$$

There are many partial transversals. A few examples are:

$$PT: p_1 = b, p_2 = a \rightarrow PT = \{a, b\}$$

$$PT: p_1 = b, p_3 = c \rightarrow PT = \{b, c\}$$

$$PT: p_1 = a, p_3 = d \rightarrow PT = \{a, d\}$$

$$PT: p_2 = a, p_3 = c \rightarrow PT = \{a, c\}$$

In this example, we have intentionally numbered the p's to coincide with the original set A.

1.8.7   EXAMPLE

Find a partial transversal of the system

$$S = \{a, b, c\}$$

$$A = \{\{a, b\}, \{a\}, \{a, b\}\}.$$

First we note that there are no transversals for this system. We must choose a from the second element. This means that we must choose b from the first element. However, the last element only allows the choices $\{a, b\}$, and we have already used both a and b.

Even though there are no transversals, there are some partial transversals. A few examples are:

$$PT: p_1 = a, p_3 = b \rightarrow PT = \{a, b\}$$

$$PT: p_1 = a \rightarrow PT = \{a\}$$

$$PT: p_1 = b \rightarrow PT = \{b\}$$

$$PT: p_1 = a, p_3 = b \rightarrow PT = \{a, b\}$$

$$PT: p_2 = a, p_3 = b \rightarrow PT = \{a, b\}$$

$$PT: p_2 = a, p_3 = b \rightarrow PT = \{a, b\}$$

$$PT: \emptyset \rightarrow PT = \emptyset$$

Here we see that there are several partial transversals for the system, even though no transversal exists.

1.8.8 PROPOSITION

Let S be a set, and P be a partial transversal of S. Any subset of P is also a partial transversal of S.

Proof. Let $R \subseteq P$ be a subset of P. If $R = P$ then R is a partial transversal because P is a partial transversal. Otherwise, $R \subset P$ which means that $P - R \neq \emptyset$.

In order to construct P, we must have some subsequence $B \subseteq A$ where the element of A are subsets $A_i \subseteq S$. Let B_i be the sequence elements of B, and let p_i be the choice made from the set B_i in constructing P.

Let $e \in P - R$. Since $e \in P$, then there must be some element k of the seuqnece B_i where P chooses e. Construct a new sequence C_i such that

$$C_i = \begin{cases} B_i & i \neq k \\ \emptyset & i = k \end{cases}.$$

Essentially, C is the same sequence as B, just with the kth element removed. Now, repeat this process for every $e \in P - R$. Each e eliminates exactly one term of B because a partial transversal cannot pick the same element twice.

The remaining sequence C is a subsequence of B, but B is a subsequence of A, so C must be a subsequence of A as well. For each of the elements left in C, choose p exactly the same way p was choosen for B.

Let $\bar{P}$ be the set of choices made from C, and let $\bar{p}_i \in \bar{P}$ be the selection made at $C_i = B_i$. We know $C \subset A$, and $\bar{P} \subseteq S$. We need to show that $\bar{P}$ and C satisfy **PT-1-4**.

PT-1: $\bar{p}_i \in C_i$ for each $\bar{p}_i \in \bar{P}$

Since $C_i = B_i$, C and B have the same slections. But the choice $\bar{p}_i$ is the same as p_i (the choice made at B_i). So $\bar{p}_i \in C_i$ for each $\bar{p}_i \in \bar{P}$.

PT-2: $|\bar{P}| = |C|$

All of the choices p_i are distinct because B is a partial transversal. So all of the choices $\bar{p}_i$ must also be distinct. We have one choice for every C_i, so it must be true that $|\bar{P}| = |C|$.

PT-3: $\bar{p}_i \neq \bar{p}_j$ for every $\bar{p}_i, \bar{p}_j \in \bar{P}$ with $i \neq j$

We know this must be true because $\bar{p}_i = p_i$, and $p_i \neq p_j$ for every $p_i, p_j \in P$ with $i \neq j$ because B is a transversal.

PT-4: $|C| < |A|$

C cannot be bigger than B. So $|C| \leq |B|$. But since B is a transversal, $|B| < |A|$. Thus, $|C| < |A|$.

From this, C, $\bar{P}$, and $\bar{p}$ meet all of the requirements of a partial transversal. Moreover, since C is a subset of P, we have shown that a subset of a partial transversal is also a partial transversal. ∎

1.8.9   PROBLEMS

1. Find all transversal of $S = \{a, b, c, d\}$ with $A = \{\{a, b, c\}, \{d\}\}$.
2. Find all transversal of $S = \{a, b, c, d\}$ with $A = \{\{a, b\}, \{c\}, \{a, b, c\}\}$.
3. Find all partial transversals for the system in problem 1.
4. Final all partial transversals for the system in problems 2.

1.9   Permutations

Permutations arise in this book when examining matroid isomorphisms. In this section we will review how permutations act on objects and the effect of successive permutations.

1.9.1   DEFINITION: PERMUTATION

> A permutation is an exchange of labels in a set.

1.9.2   EXAMPLE

Let S be the set $\{a, b\}$. Apply the permutation $a \leftrightarrow b$.

The permutation exchanges the labels a and b everywhere. The relabled (permuted) set is

$$S' = PS = \{b, a\}$$

However, the order of elements in a set is immaterial. So

$$S' = \{b, a\} = \{a, b\} = S$$

Any relabeling of a set will produce the same set.

1.9.3   EXAMPLE

Let $S = \{a, b\}$ be a set and let $f: S \times S$ be the map

$$
\begin{aligned}
(a, a) &= 0 \quad (a, b) = 1 \\
(b, a) &= -1 \quad (b, b) = 0
\end{aligned}
$$

Apply the permutation $(a\ b)$ to the map.

We can generally designate a permutation using the notation $(a\ b\ c)$ to mean that we relabel a with b, b with c, and c with a. In this case, $(a\ b)$ means we swap a and b everywhere.

Example 1.9.2 shows that when we apply the permutation to S, we get S back. Next, apply the permutation to f:

$$
f': \begin{aligned}
(b, b) &= 0 \quad (b, a) = 1 \\
(a, b) &= -1 \quad (a, a) = 0
\end{aligned}
$$

Rearranging,

$$
f': \begin{aligned}
(a, a) &= 0 \quad (a, b) = -1 \\
(b, a) &= 1 \quad (b, b) = 0
\end{aligned}
$$

We see that although the set S is not changed under the permutation, the map f is different.

1.9.4   PROPERTIES

We may need to apply two permutations successively. The successive application of two permutations is the same as a single application of another permutation.

We use a different permutation notation to determine the combined result. The permutation $(a\ b)$ may also be written as

$$\begin{pmatrix} a & b \\ b & a \end{pmatrix}$$

In this notation, the top row is the original set, and the bottom row is the replacement.

To examine the result of successive permutations, we place the permutations side by side in order with the leftmost permutation the first we want to apply, then trace through the replacements.

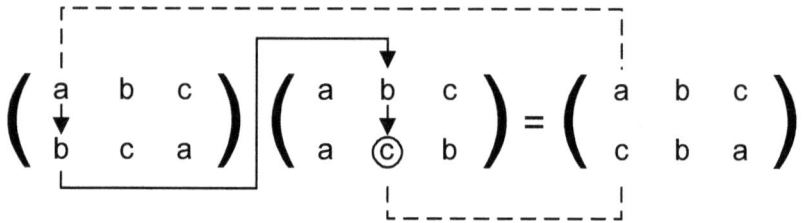

Figure 4: Example of application of multiple permutations as a single permutation.

Figure 4 provides an example of applying two permutations. To determine the overall effect, we start form an initial element, see what it becomes after the permutation, then lookup that element on the next permutation, see what that becomes, and repeat until the end. We repeat this process for each element to determine the final mapping.

Chapter 2: Matroid Classifications

Matroids generalize the concept of linear independence. Linear independence appears throughout mathematics. Because of this, there are many different applications that give rise to matroids. This chapter examines the most common sources of matroids.

2.1 Vector Matroids

Vector matroids are one of two fundamental classes of matroids. This section introduces vector matroids and establishes an axiomatic definition for vector matroids.

2.1.1 DEFINITION: SET MATROID

> Let S be a finite set (ground set). Let T be a collection of subsets of S with the following properties:
>
> **SM-1** $\emptyset \in T$
>
> **SM-2** If $t \in T$ and $\bar{t} \subseteq t$, then $\bar{t} \in T$
>
> **SM-3** Let $t_1, t_2 \in T$ where $|t_1| < |t_2|$, then $\exists\, d \in t_2 - t_1$ such that $t_1 \cup d \in T$
>
> Then the ordered pair (S, T) is called a matroid.

In the above definition, a matroid is formed from a set and a collection of its subsets. The collection of subsets T is called the independent set of the matroid. A matroid based on the above definition will be termed a set matroid.

Later, we show this set-based definition is equivalent to a circuit-based definition. In both cases, the resulting structure is a matroid. However, the two structures arise from different contexts, and their equivalence is not immediately apparent.

2.1.2 EXAMPLE

Let $S = \{\emptyset\}$ and $T = \{\emptyset\}$. Show (S, T) forms a matroid.

Examine each axiom:

SM-1

We have $\emptyset \in T$ by definition of T, so **SM-1** is satisfied.

SM-2

T has only one element, $\emptyset$. However, there are no other distinct subsets of $\emptyset$. Looking at **SM-2**, we have examined every t and $\bar{t}$, and determined that in every case where $t \in T$ and $\bar{t} \subseteq t$, then $\bar{t} \in T$. Thus, **SM-2** is satisfied.

SM-3

There is only one element of T, so there are no two elements $t_1, t_2 \in T$ where $|t_1| < |t_2|$. Thus, **SM-3** is trivially satisfied.

2.1.3   EXAMPLE

Let $S = \{a\}$ and $T = \{\emptyset\}$. Show (S, T) forms a matroid.

Examine each axiom:

SM-1

We have $\emptyset \in T$ by definition of T, so **SM-1** is satisfied.

SM-2

T has only one element, $\emptyset$. Again, there are no other distinct subsets of $\emptyset$. Examining **SM-2**, we have reviewed every t and $\bar{t}$, and determined that in every case where $t \in T$ and $\bar{t} \subseteq t$, then $\bar{t} \in T$. Thus, **SM-2** is satisfied.

SM-3

There is only one element of T, so there are no two elements $t_1, t_2 \in T$ where $|t_1| < |t_2|$. Thus, **SM-3** is trivially satisfied.

2.1.4   EXAMPLE

Let $S = \{a\}$ and $T = \{\emptyset, \{a\}\}$. Show (S, T) forms a matroid.

Examining each axiom:

SM-1

We have $\emptyset \in T$ by definition of T, so **SM-1** is satisfied.

SM-2

T has two elements, $\emptyset$ and $\{a\}$. For $\emptyset$, there are no other distinct subsets to review.

For $\{a\}$, there are two subsets, $\emptyset$ and $\{a\}$. Both of these subsets are elements of T.

In each case, if $t \in T$ and $\bar{t} \subseteq t$, then $\bar{t} \in T$, so **SM-2** is satisfied.

SM-3

There are two elements of T. Set $t_1 = \emptyset$ and $t_2 = \{\{a\}\}$. We need to identify a set $d \in t_2 - t_1$ such that $t_1 \cup d \in T$.

First, $t_2 - t_1 = \{\{a\}\}$. This set only has one member, so is must be that $d = \{a\}$. Checking **SM-3**,

$$t_1 \cup d = \{\{a\}\} \in T$$

so **SM-3** is satisfied.

2.1.5   DEFINITION: VECTOR MATROID

Let A be an $m \times n$ matrix. Label each of the n columns with a unique label λ_i, and form a set Λ from the column labels. Let I be the set of all sets of column labels where the matrix columns are linearly independent. Then (Λ, I) is a matroid.

The definition and following proof may be somewhat confusing. The examples following this section may shed light on the definition and the elements of the proof below. If the first examination of the proof is unclear, the reader is encouraged to review the examples and return to the proof.

Proof. Examine each of the three set matrix axioms.

SM-1

The empty set of column vectors is trivially a set of linearly independent columns. So $\emptyset \in T$, satisfying **SM-1**.

SM-2

Let $t \in I$ and set $p = |t|$. The elements of t are the column labels $\lambda_k \in t$. By definition the matrix columns associated with these labels are linearly independent.

Let v_k represent the matrix column associated with the column header λ_k. Linear independence of the v_k's means if

$$a_1 v_1 + a_2 v_2 + \cdots + a_p v_p = 0$$

then each $a_k = 0$.

Examine a subset $\bar{t} \subseteq t$. The v_k's corresponding to the column headers of $\bar{t}$ must be linearly independent because any subset of a linearly independent set is also linearly independent.

However, I is the set of all sets of column labels that correspond to a linearly independent set of column vectors. Since the column headers in $\bar{t}$ form a linear independent set, $\bar{t} \in I$.

We have for each $t \in I$ and $\bar{t} \subseteq t$, then $\bar{t} \in I$, so **SM-2** is satisfied.

SM-3

Choose two elements of I, $t_1, t_2 \in I$ where $|t_1| < |t_2|$. Both t_1 and t_2 correspond to linearly independent sets of column vectors. Now, since $|t_1| < |t_2|$, there are fewer vectors in t_1 than in t_2. Thus, the dimension of the vector space formed from the t_1 vectors is smaller than the dimension of the vector space formed from the t_2 vectors. Therefore, there must be at least one vector in t_2 that is not in t_1, and this vector must have a component orthogonal to all of the vectors in t_1.

Let $\bar{v}$ be a vector corresponding to a column header in t_2 that has a component orthogonal to all of the vectors from t_1. Let $\bar{\lambda} \in t_2$ be the column header corresponding to $\bar{v}$. Now, because $\bar{v}$ has a component orthogonal to all of the vectors from t_1, $\bar{\lambda} \in t_2 - t_1$. Otherwise $\bar{\lambda}$ would be a member of t_1 and then $\bar{v}$ would not have a component orthogonal to all vectors from t_1.

Let v_k be the vectors from t_1. Orthogonality conditions require

$$a_1 v_1 + a_2 v_2 + \cdots + a_k v_k + \bar{a}\bar{v} = 0$$

only when $\bar{a} = 0$.

But since the v_k's are linearly independent, the independence relation

$$a_1 v_1 + a_2 v_2 + \cdots + a_k v_k = 0$$

insists that each $a_k = 0$.

Putting these two conditions together, if

$$a_1 v_1 + a_2 v_2 + \cdots + a_k v_k + \bar{a}\bar{v} = 0$$

then each $a_k = 0$ and $\bar{a} = 0$, which means that $\bar{v}$ is linearly independent of the v_k's.

Let $\bar{\lambda} \in t_2 - t_1$ be the column header corresponding to $\bar{v}$. Then the column vectors corresponding to $t_1 \cup \bar{\lambda}$ are linearly independent. But since I is the set of all sets of column labels where the matrix columns are linearly independent, then $t_1 \cup d \in I$.■

This definition of a vector matroid may be easier to understand by means of examples. The next several examples are intended to bring clarity to the definition.

2.1.6   EXAMPLE

Form a matroid from the 1×1 matrix $[1]$.

First, label the column headers. There is only one column, which we label as column a. The set of column labels is $\Lambda = \{a\}$.

Next, form the set of all column headers that correspond to a linearly independent set of column vectors. Clearly the empty set is a trivial example, and the set $\{a\}$ forms a linearly independent set as well.

The set of all sets of column labels where the matrix columns are linearly independent is $I = \{\emptyset, \{a\}\}$. The ordered pair (Λ, I) is a vector matroid. This is the same as the matroid of Example 2.1.4.

2.1.7   EXAMPLE

Find the vector matroid from the 2×3 matrix $\begin{bmatrix} 1 & 0 & 1 \\ 0 & 1 & 1 \end{bmatrix}$.

Label the columns a, b and c respectively; $\Lambda = \{a, b, c\}$ with corresponding vectors $v_a = \begin{pmatrix} 1 \\ 0 \end{pmatrix}$, $v_b = \begin{pmatrix} 0 \\ 1 \end{pmatrix}$ and $v_c = \begin{pmatrix} 1 \\ 1 \end{pmatrix}$. Each pair of vectors is linearly independent, as is each individually. Furthermore, the empty set is another linearly independent set of column headers. The set of all sets of linearly independent column headers is

$$I = \{\emptyset, \{a\}, \{b\}, \{c\}, \{a, b\}, \{a, c\}, \{b, c\}\}.$$

The ordered pair (Λ, I) is a vector matroid.

2.1.8   EXAMPLE

Find the vector matroid from the 2×3 matrix $\begin{bmatrix} 1 & 0 & 0 \\ 0 & 1 & 0 \end{bmatrix}$.

Label the columns a, b and c respectively; $\Lambda = \{a, b, c\}$ with corresponding vectors $v_a = \begin{pmatrix} 1 \\ 0 \end{pmatrix}$, $v_b = \begin{pmatrix} 0 \\ 1 \end{pmatrix}$ and $v_c = \begin{pmatrix} 0 \\ 0 \end{pmatrix}$. Unlike the previous example, v_c is not linearly independent from the other vectors. In fact, v_c is not even linearly independent when taken by itself. Specifically, in the equation for linear independence

$$c \begin{pmatrix} 0 \\ 0 \end{pmatrix} = \begin{pmatrix} 0 \\ 0 \end{pmatrix}$$

The coefficient c does not need to be 0. Thus, the set $\{c\}$ is not a set of linearly independent column headers.

However, v_a and v_b are linearly independent because the equation

$$a \begin{pmatrix} 1 \\ 0 \end{pmatrix} + b \begin{pmatrix} 0 \\ 1 \end{pmatrix} = \begin{pmatrix} 0 \\ 0 \end{pmatrix}$$

means that both a and b must be 0. Similarly, if we have either of the equations

$$a \begin{pmatrix} 1 \\ 0 \end{pmatrix} = \begin{pmatrix} 0 \\ 0 \end{pmatrix} \quad b \begin{pmatrix} 0 \\ 1 \end{pmatrix} = \begin{pmatrix} 0 \\ 0 \end{pmatrix}$$

then $a = 0$ and $b = 0$.

In this example $I = \{\emptyset, \{a\}, \{b\}, \{a, b\}\}$, and the ordered pair (Λ, I) is a vector matroid.

2.1.9 DEFINITION: RESTRICTION MATROID

> Let $M = (S, T)$ be a matroid. Let $X \subseteq S$. The independent sets $\bar{T} = T|X$ are the sets in T that are subsets of X. The set $(X, \bar{T})$ forms a matroid called a restriction matroid.

Proof. Show that each of the matroid axioms are satisfied.

SM-1

$\emptyset \in T$ because $M = (S, T)$ is a matroid and the independent set of a matroid must contain $\emptyset$.

But for any set X, $\emptyset \subset X$. Thus, the set $\emptyset$ is a set in T that is also a subset of X, which means that $\emptyset \in \bar{T}$.

SM-2

Let $t \in \bar{T}$ and $\bar{t} \subseteq t$. If $t \in \bar{T}$ then the definition of $\bar{T}$ means that t must be a subset of X. Since $\bar{t} \subseteq t$, and $t \subseteq X$, then $\bar{t} \subseteq X$. Also, $\bar{t} \in T$ because $t \in T$ and every subset of an independent set is also an independent set by **SM-2**. However, if $\bar{t} \in T$ and $\bar{t} \subseteq X$, then $\bar{t} \in \bar{T}$.

SM-3

Let $t_1, t_2 \in \bar{T}$ with $|t_1| < |t_2|$. But $t_1, t_2 \in T$. So $\exists d \in t_2 - t_1$ such that $t_1 \cup d \in T$. But since $t_1, t_2 \subseteq X$, is must be true that $t_2 - t_1 \subseteq X$. Therefore, $t_1 \cup d \subseteq X$, which means $t_1 \cup d \in \bar{T}$, satisfying **SM-3**. ∎

2.1.10 PROBLEMS

1. Let $S = \{a, b\}$ and $T = \{\emptyset\}$. Show (S, T) forms a matroid.
2. Let $S = \{a, b\}$ and $T = \{\emptyset, \{a\}\}$. Show (S, T) forms a matroid.
3. Let $S = \{a, b\}$ and $T = \{\emptyset, \{a\}, \{b\}\{a, b\}\}$. Show (S, T) forms a matroid.
4. Let $S = \{a, b\}$ and $T = \{\{a, b\}\}$. Show (S, T) is not a matroid.
5. Let $S = \{a, b\}$ and $T = \{\emptyset, \{a\}, \{a, b\}\}$. Show (S, T) is not a matroid.
6. Let $S = \{a, b\}$ and $T = \{\emptyset, \{a, b\}\}$. Show (S, T) is not a matroid.
7. Find the vector matroid from the matrix $\begin{bmatrix} 1 & 0 & 1 & 0 \\ 0 & 1 & 1 & 1 \end{bmatrix}$.
8. Find the vector matroid from the matrix $\begin{bmatrix} 1 & 0 & 0 & 0 & 1 \\ 0 & 1 & 0 & 1 & 1 \\ 0 & 0 & 1 & 1 & 0 \end{bmatrix}$.
9. Let V be the matroid from problem 7. Find the restriction matroid $R = V|\{a, b\}$.
10. Let V be the matroid from problem 8. Find the restriction matroid $R = V|\{a, b, c\}$.

2.2   Circuits

Matroids contain a set of independent sets. A set of dependent sets may be formed as well. Circuits are the set of minimally dependant sets of the set of dependant sets of a matroid.

2.2.1 DEFINITION: CIRCUIT

Let M be a matroid (Λ, I). Let P be the power set of Λ. The set of dependent sets is $D = P - I$. A circuit is an element $c \in \lfloor D \rfloor$.

2.2.2 EXAMPLE

Find the circuits of the matroid (Λ, I) where $\Lambda = \{a, b, c\}$ and $I = \{\emptyset, \{a\}, \{b\}, \{c\}, \{a, b\}, \{a, c\}, \{b, c\}\}$.

The power set of Λ is

$$P = \{\emptyset, \{a\}, \{b\}, \{c\}, \{a, b\}, \{a, c\}, \{b, c\}, \{a, b, c\}\}.$$

From the power set, the set of dependent sets is

$$D = P - I = \{\{a, b, c\}\}.$$

The minimally dependent set is

$$\lfloor D \rfloor = \{\{a, b, c\}\}.$$

2.2.3 EXAMPLE

Find the circuits of the matroid (Λ, I) where $\Lambda = \{a, b, c\}$ and $I = \{\emptyset, \{a\}, \{b\}, \{a, b\}\}$.

The power set of Λ is

$$P = \{\emptyset, \{a\}, \{b\}, \{c\}, \{a, b\}, \{a, c\}, \{b, c\}, \{a, b, c\}\}.$$

From the power set, the set of dependent sets is

$$D = P - I = \{\{c\}, \{a, c\}, \{b, c\}, \{a, b, c\}\}.$$

The circuits of the matroid are the minimal elements

$$\lfloor D \rfloor = \{\{c\}\}.$$

2.2.4 THEOREM

Let M be a matroid (Λ, I), and let P be the power set of Λ. For every element $p \in P$, either $p \in I$, or there exists a circuit c where $c \subseteq p$.

Proof. Let $D = P - I$. From the definition of D, every element of P is either a member of D or I. Furthermore, $D \cap I = \emptyset$, so no element of P is in both D and I.

However, for every element $d \in D$, there is no element $i \subset I$ where $d \subseteq i$. By **SM-2**, if $d \subseteq i$, then $d \subset I$, contradicting the assertion that $d \in D$.

The circuits of M are minimally dependant members of D. Every minimally dependent member of D must be a subset of some element in D.

Thus, every element $p \in P$ satisfies either

(i) $p \in I$; or

(ii) $\exists\, c \in C \mid c \subseteq P.$ ∎

2.2.5 THEOREM

Let C be the set of circuits of a matroid. Then $\emptyset \notin C$.

Proof. Let M be the matroid with I the set of independent sets. The set of circuits is $C = \lfloor D \rfloor = \lfloor P - I \rfloor$. But $\emptyset \in P$ and $\emptyset \in I$ so $\emptyset \notin \lfloor P - I \rfloor$, which means $\emptyset \notin C$. ∎

2.2.6 THEOREM

Let C be the set of circuits of a matroid, and let $c_1, c_2 \in C$. If $c_1 \subseteq c_2$, then $c_1 = c_2$.

Proof. Let M be the matroid with I the set of independent sets. The set of circuits is $C = \lfloor D \rfloor = \lfloor P - I \rfloor$. Let $c_1, c_2 \in C$. The elements of C are minimally dependent members of $P - I$. By definition of minimal members of a set, $c_1 \not\subset c_2$. Thus, if $c_1 \subseteq c_2$, then $c_1 = c_2$. ∎

2.2.7 THEOREM

Let c_1, c_2 be distinct circuits of a matroid, and let $e \in c_1 \cap c_2$. Then there is a circuit $c_3 \subseteq (c_1 \cup c_2) - e$.

Proof. First, $c_1 \cap c_2 \neq \emptyset$ because c_1, c_2 are distinct, and by Theorem 2.2.6, if one is a subset of the other, the two must be equal.

If we assume $(c_1 \cup c_2) - e$ does not contain a circuit, then $(c_1 \cup c_2) - e$ must be independent. All subsets of $(c_1 \cup c_2) - e$ are also independent by property **SM-2**.

Let $f \in c_2 - c_1$. Choose a set $I \subseteq c_1 \cup c_2$ that is a maximal independent set with $c_2 - f \subseteq I$. Such a set must exist because $c_2 - f$ is an independent set, and $c_2 - f \subseteq c_1 \cup c_2$. However, $f \notin I$, otherwise $c_2 \subseteq I$, which would mean c_2 is independent by **SM-2**, contradicting the assertion c_2 is a circuit.

Similarly, there must be some element $g \in c_1$ such that $g \notin I$, by the same reasoning as above. Hence,

$$|I| \leq |(c_1 \cup c_2) - \{f,g\}| = |(c_1 \cup c_2)| - 2 < (c_1 \cup c_2) - \{f,g\}$$

Applying **SM-3** with $t_1 = I$ and $t_2 = (c_1 \cup c_2) - \{f,g\}$, there exists an independent set $t_3 = I \cup h$, where $h \subset (c_1 \cup c_2) - \{f,g\} - I$. But this would mean $|t_3| > |I|$ while $I \subset t_3$.

This contradicts the assumption that I is a maximal independent set. Therefore, $(c_1 \cup c_2) - e$ must not be an independent set, and from Theorem 2.2.4, there must be some circuit $c_3 \subseteq (c_1 \cup c_2) - e$.∎

2.2.8 DEFINITION: CIRCUIT MATROID

> Let S be a finite set. Let C be a collection of subsets of S with the following properties:
>
> **CM-1** $\emptyset \notin C$
>
> **CM-2** If $c_1, c_2 \in C$ and $c_1 \subseteq c_2$, then $c_1 = c_2$
>
> **CM-3** If c_1, c_2 are distinct members of C and $e \in c_1 \cap c_2$, then there exists a $c_3 \in C$ such that $c_3 \subseteq (c_1 \cup c_2) - e$
>
> Let $\mathfrak{I}$ be the collection of subsets of S that do not contain any member of C as a subset. Then the ordered pair $(S, \mathfrak{I})$ is a matroid.

In the above definition, a matroid is formed from a set and a collection of circuits. A matroid based on the above definition will be termed a circuit matroid.

Next, we show this circuit-based definition is equivalent to the set-based definition provided earlier.

Proof. Let C be a collection of subsets of S satisfying **CM-1-3**. The power set P of S connects these circuits to a set matroid. We need to show that the set $\mathfrak{I}$ satisfies **SM-1-3**.

SM-1 $\emptyset \in \mathfrak{I}$

From **CM-1**, $\emptyset \notin C$. The set $\emptyset$ only has $\emptyset$ as a subset. Since $\emptyset \notin C$, and from the definition of $\mathfrak{I}$, $\emptyset \in \mathfrak{I}$.

SM-2   If $t \in \mathfrak{I}$ and $\bar{t} \subseteq t$, then $\bar{t} \in \mathfrak{I}$

If $t \in \mathfrak{I}$, then t does not contain any member of C as a subset. Moreover, no subset $\bar{t} \subseteq t$ may contain a member of C as a subset, otherwise t would contain this element as well. Therefore, since $\bar{t}$ does not contain any element of C as a subset, $\bar{t} \in \mathfrak{I}$.

SM-3   Let $t_1, t_2 \in \mathfrak{I}$ where $|t_1| < |t_2|$, then $\exists\, d \in t_2 - t_1$ such that $t_1 \cup d \in \mathfrak{I}$

Assume that **SM-3** does not hold, and let $t_1, t_2 \in \mathfrak{I}$ where $|t_1| < |t_2|$. Let $d \in t_2 - t_1$. $t_1 \not\subset t_2$ because in that case $t_1 \cup d \subset t_2$ and by **SM-2**, every subset of t_2 is a member of $\mathfrak{I}$.

Choose $t_3 \subseteq t_1 \cup t_2$ such that $|t_3| > |t_1|$, $I_3 \in \mathfrak{I}$, and $|t_3 - t_1|$ is a minimum. Since $t_3 = t_2$ satisfies the first and second criteria, $t_3 \neq \emptyset$. Moreover, $t_1 - t_3 \neq \emptyset$, otherwise we would have $t_1 \subset t_3$ and every element of $f \in t_3 - t_1$ would satisfy $t_1 \cup f \in \mathfrak{I}$ because every subset of $t_3 \in \mathfrak{I}$ by **SM-2**.

Examine $e \in t_1 - t_3$ and let $T_f = (t_3 \cup e) - f$. Because $|t_3 - t_1|$ is a minimum, $T_f \notin \mathfrak{I}$. Thus, there exists some $c_f \in C$ such that $c_f \subseteq T_f$. From the definition of T_f, $f \notin c_f$, but $e \in c_f$ because if $e \notin c_f$ then $c_f \subseteq t_3$ which would mean that $t_3 \notin \mathfrak{I}$ contrary to the assumption.

Let $g \in t_3 - t_1$. We know that $C_g \cap (I_3 - I_1) \neq \emptyset$, otherwise we would have $C_g \subseteq (I_3 \cap I_1) \cup e - g \subseteq I_1$ which cannot be true because $I_1 \in \mathfrak{I}$. Let $g \in C_g \cap (I_3 - I_1)$. We know that $e \in C_g \cap C_h$, so we can use **CM-3** to arrive at some $c_3 \in C$ such that $c_3 \subseteq (c_g \cup c_h) - e$. However, we see that C_g and C_h are both subsets of $I_3 \cup e$. This means that $c_3 \subseteq I_3$ which is not possible from the assumption that $I_3 \in \mathfrak{I}$. Thus, the initial assumption that **SM-3** does not hold is invalid. ∎

2.3   Graphic Matroids

Matroids arise from graph theory by associating the cycles of a graph with the circuits of a matroid. This section introduces the concepts allowing the association of graphs with matroids.

2.3.1   DEFINITION: CYCLE MATROID

Let G be a graph. Label each edge in G, and let E be the set of edges. Let Γ be the set of cycles in G. The ordered pair (E, Γ) forms a cycle matroid.

The cycle matroid is a particular instance of a circuit matroid as demonstrated below.

CM-1 $\emptyset \notin C$

Every cycle of a graph must contain some set of edges, so $\emptyset \notin C$.

CM-2   If $c_1, c_2 \in C$ and $c_1 \subseteq c_2$, then $c_1 = c_2$

A graph cycle cannot contain another cycle. So if $c_1, c_2 \in C$ and $c_1 \subseteq c_2$, then the two cycles must be the same, i.e. $c_1 = c_2$.

CM-3   If c_1, c_2 are distinct members of C and $e \in c_1 \cap c_2$, then there exists a $c_3 \in C$ such that $c_3 \subseteq (c_1 \cup c_2) - e$

If two distinct cycles have $c_1 \cap c_2 \neq \emptyset$, then the cycles must have at least one edge in common. In this case, there must be another cycle containing these cycles with an edge removed.

Figure 5 shows a pair of cycles with a common edge and another cycle $c_3 \subseteq (c_1 \cup c_2) - e$. The cycle $\{a, b, c, d, e, f\}$ and the cycle $\{a', b', c', d', e', f', f\}$ have the edge f in common between them. However, there is also a cycle $\{a, b, c, d, e, a', b', c', d', e', f'\}$ that has all of the edges of both cycles, minus the common edge f. ■

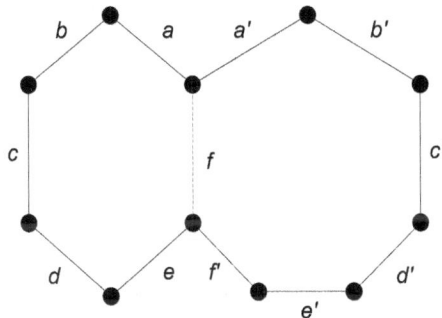

Figure 5: Graph cycles.

2.3.2   DEFINITION: GRAPHIC MATROID

> Let M be a matroid. If M is isomorphic to the cycle matroid of a graph, then M is a graphic matroid.

2.3.3   EXAMPLE

Find the matroid from the graph in Figure 6.

The edge set of the graph is $E = \{a, b, c\}$, and there is one cycle in the graph $\Gamma = \{\{a, b, c\}\}$. The ordered pair (E, Γ) forms a cycle matroid.

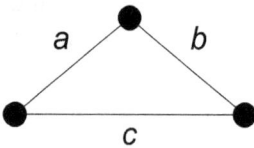

2.3.4   DEFINITION: PARALLEL

> Let $M = (E, \Gamma)$ be a graphic matroid. Let $\gamma \in \Gamma$ such that $|\gamma| = 2$. The element of γ are called parallel in M.

2.3.5   EXAMPLE

Find the matroid from the graph in Figure 7.

This graph has a set of 'parallel' edges. The edge set of the graph is $E = \{a, b\}$, and there is one cycle in the graph $\Gamma = \{\{a, b\}\}$. The ordered pair (E, Γ) forms a cycle matroid.

Figure 6: Three edge graph.

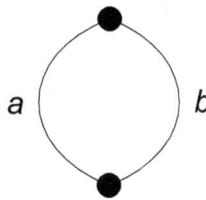

Figure 7: Two edge graph.

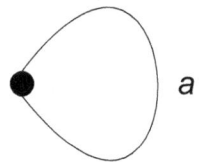

Figure 8: Single edge graph.

If circuit of a matroid has exactly two elements, say $\{a, b\}$, then the elements a, b are called parallel. The matroid from graph in Figure 7 is the simplest graphic matroid with parallel elements.

2.3.6   DEFINITION: LOOP

> Let $M = (E, \Gamma)$ be a graphic matroid. A loop is an element $\gamma \in \Gamma$ such that $|\gamma| = 1$.

2.3.7   EXAMPLE

Find the matroid from the graph in Figure 8.

The graph of Figure 8 has a loop. The edge set of the graph is $E = \{a\}$, and there is one cycle in the graph $\Gamma = \{\{a\}\}$. The ordered pair (E, Γ) forms a cycle matroid.

A loop is a graph cycle consisting of only a single edge. The graph of Figure 8 is the simplest graph with a loop.

2.3.8   EXAMPLE

Find the matroid from the graph in Figure 9.

This graph is more complicated and has three cycles. The edge set of the graph is $E = \{a, b, c, d, e, f, a', b', c', d', e', f'\}$. The cycle set is

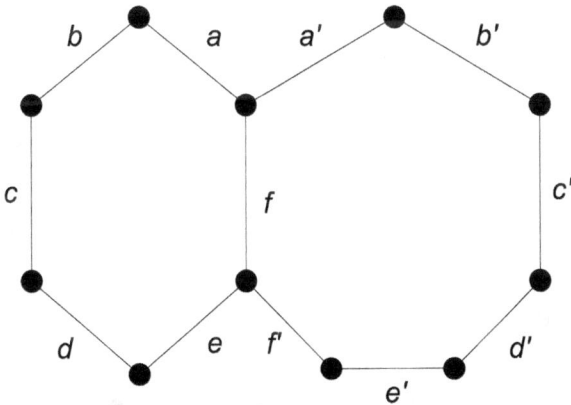

Figure 9: Three cycle graph.

$\Gamma =$
$\{\{a, b, c, d, e, f\}, \{a', b', c', d', e', f', f\}, \{a, b, c, d, e, a', b', c', d', e', f'\}\}$

There are three cycles that represent the three possible paths starting from a vertex, traversing the graph, and returning to the same vertex. The number of cycles increases rapidly with the number of faces sharing an edge. The next example demonstrates how the number of cycles increases by simply adding one edge to this graph.

2.3.9   EXAMPLE

Find the matroid from the graph in Figure 10.

The graph of Figure 10 is identical to the graph in Figure 9, with one additional edge.

The edge set is $E = \{a, b, c, d, e, f, g, a', b', c', d', e', f'\}$, which is only one more than the previous example. However, where the previous graph had three cycles, the graph of Figure 10 has six cycles. The cycles are:

$$\{a, b, c, d, e, f\} \quad \{a', b', c', d', e', f', f\} \quad \{a, b, c, d, e, a', b', c', d', e', f'\}$$
$$\{a, b, g\} \quad \{c, d, e, f, g\} \quad \{a', b', c', d', e', f', e, d, c, g\}$$

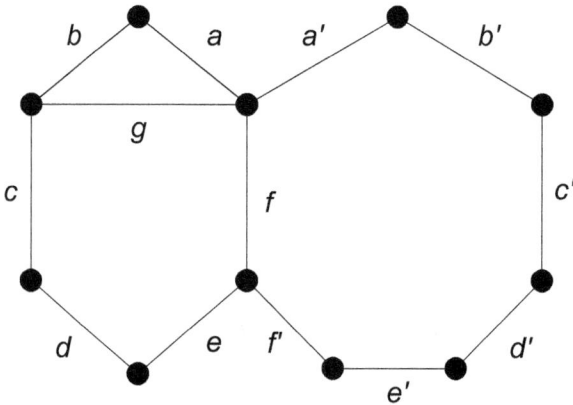

Figure 10: Six cycle graph.

2.3.10 DEFINITION: PARALLEL ELEMENTS

> Let $M = (S, \Gamma)$ be a matroid. Let $\gamma \in \Gamma$ such that $|\gamma| = 2$. The two elements of γ are called parallel in M.

In a graph, parallel elements are a pair of edges connecting the same vertices.

2.3.11 DEFINITION: PARALLEL CLASS

> Let $M = (S, T)$ be a matroid. Let $P \subseteq S$ be a maximal set such that any pair of elements in P are parallel in M. The maximal set P is a parallel class in M.

Looking at graphs, parallel elements are pairs of edges connecting the same pair of vertices. A parallel class is the set of all edges connecting a pair of vertices.

You can have a parallel class containing only one edge. This corresponds to the case where only one edge connects the vertices in question. In this case, the parallel class is called trivial.

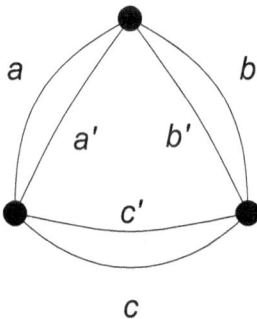

Figure 11: Three vertex graph with parallel edges.

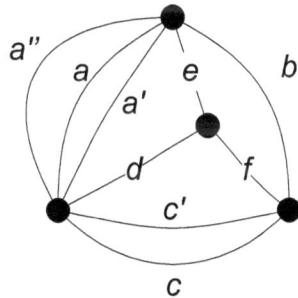

Figure 12: Four edge graph with parallel edges.

2.3.12 EXAMPLE

Find the non-trivial parallel classes for the diagram in Figure 11.

There are three cycles with only two elements:

$$\{a, a'\} \quad \{b, b'\} \quad \{c, c'\}$$

These are the parallel elements for graphic matroid arising from this diagram. These are also the maximal sets for each of the parallel elements, making these the non-trivial parallel classes. There are no trivial parallel classes in this example, as there are no pair of vertices that are connected by a single edge.

2.3.13 EXAMPLE

Find the non-trivial parallel elements for the diagram in Figure 12**Error! Reference source not found.**.

There are four cycles with only two elements:

$$\{a, a'\} \quad \{a, a''\} \quad \{a', a''\} \quad \{c, c'\}$$

So the graphic matroid arising from this diagram has four parallel elements. The first three can be combined together into the parallel class $M_a = \{a, a', a''\}$. Any two distinct elements $m_i \in M_a$ are present as parallel elements in the matroid. Since $|M_a| = 3$, the parallel class is non-trivial.

There are four trivial parallel classes in this example:

$$\{b\} \quad \{d\} \quad \{e\} \quad \{f\}.$$

2.3.14 DEFINITION: SIMPLE MATROID

Let M be a graphic matroid. If M has no loops and no non-trivial parallel classes, M is called a simple matroid (combinatorial geometry).

2.3.15 EXAMPLE

Is the graphic matroid from Example 2.3.9 simple?

The graphic matroid from Example 2.3.9 is a simple matroid. Looking at the list of cycles of the matroid, there are no cycles with dimension 1 or 2.

2.3.16 EXAMPLE

Are the graphic matroids from Examples 2.3.12 and 2.3.13 simple?

The graphic matroid from example 2.3.12 is not simple. The graph has no loops, but three parallel elements.

The graphic matroid from Example 2.3.13 is also not simple. Again, there are no loops, but there are two non-trivial parallel classes.

2.4   Uniform Matroids

Uniform matroids are a class of matroids formed by specifying the bases of the matroid. The bases are the set of maximal independent sets and are discussed in detail in section 3.1 which contains the definition of the Basis Matroid used in this section. This class of matroids is linked to combinatorial analysis as the bases are chosen as the m element subsets of an n element set.

2.4.1   DEFINITION: UNIFORM MATROID

> Let S be a set where $|S| = n$. Let U be the collection of all m element subsets of S. The ordered pair (S, U) forms a uniform matroid is U is taken as the set of bases of M. Uniform matroids are designated $U_{m,n}$.

In section 3.1 we show that demonstrated that a matroid may be formed from a set and a collection of subsets satisfying **BM-1** and **BM-2**:

BM-1 $B \neq \emptyset$

BM-2   For every distinct $B_1, B_2 \in B$, and every element $b \in B_1 - B_2$, there exists an element $\bar{b} \in B_2 - B_1$ such that $(B_1 - b) \cup \bar{b} \in B$

In order to show the definition of a uniform matroid satisfies Definition 2.4.1, we only need to show that the collection of subsets satisfies the basis relations.

Proof.

We show that U satisfies both **BM-1** and **BM-2**.

BM-1

The collection of m-element subsets of S always has at least one element. In the case of $m = 0$, the zero element subset is $\emptyset$. Consequently, $|U| > 0$ for every m.

BM-2

Let U_1 and U_2 be distinct members of U. Both members have the same number of element, $|U_1| = |U_2| = m$. Since they are distinct, there must be at least one element in U_1 that is not in U_2. Similarly, there must be at least one element in U_2 that is not in U_1.

Let $u_1 \in U_1 - U_2$ and $u_2 \in U_2 - U_1$. Start with the set U_1 and remove u_1, then add u_2: $\bar{U} = (U_1 - u_1) \cup u_2$. But $|\bar{U}| = m$, because $\bar{U}$ was constructed by removing a unique element from U_1 and replacing it with a unique element in U_2.

Since $|\bar{U}| = m$, $\bar{U} \in U$ because U is the collection of all m-element subsets of S. So **BM-2** is satisfied. ∎

2.4.2 LEMMA

For a uniform matroid $U_{m,n}$, the dimension of the set of bases of $U_{m,n}$ is $|B| = \binom{n}{m}$.

Proof. This is a straightforward result from the definition of a uniform matroid. The set of bases are the m-element subsets of an n-element set. The number of distinct subsets of an n-element set is given by the combinatorial $\binom{n}{m}$. Each of these subsets is a unique basis for $U_{m,n}$, hence $|B| = \binom{n}{m}$. ∎

2.4.3 EXAMPLE

Find the four uniform matroids from the set $S = \{a, b, c\}$.

0-Elements

The basis is $U = \{\emptyset\}$, and the independent set is $I = \{\emptyset\}$.

This is the matroid $U_{0,3}$.

1-Elements

The basis is $U = \{\{a\}, \{b\}, \{c\}\}$, and the independent sets are $I = \{\emptyset, \{a\}, \{b\}, \{c\}\}$.

This is the matroid $U_{1,3}$.

2-Elements

The basis is $U = \{\{a,b\}, \{a,c\}, \{b,c\}\}$, and the independents are $I = \{\emptyset, \{a\}, \{b\}, \{c\}, \{\{a,b\}, \{a,c\}, \{b,c\}\}\}$.

This is the matroid $U_{2,3}$.

3-Elements

The basis is $U = \{\{a,b,c\}\}$, and the independents are $I = \{\emptyset, \{a\}, \{b\}, \{c\}, \{a,b\}, \{a,c\}, \{b,c\}, \{a,b,c\}\}$.

This is the matroid $U_{3,3}$.

2.4.4 EXAMPLE

Find the basis for the uniform matroid $U_{4,5}$.

Let S be the set of the matroid. For $U_{4,5}$, $|S| = 5$. Designate the elements of $S = \{a, b, c, d, e\}$. The bases for a uniform matroid are the 4-element subset of S. Specifically,

$$U = \{\{a,b,c,d\}, \{a,b,c,e\}, \{a,b,d,e\}, \{a,c,d,e\}, \{b,c,d,e\}\}.$$

2.4.5 DEFINITION: EMPTY MATROID

> The unique uniform matroid $U_{0,0}$ is the matroid on the empty set and is called the empty matroid.

2.5 Paving Matroids

A paving matroid is a special class of matroids where the every dependent set has cardinality greater than or equal to the rank. Rank is discussed in detail in section 3.1. This section uses only elementary rank concepts.

2.5.1   DEFINITION: PAVING MATROID

> Let M be a matroid (S, T), and let $r(M)$ be the rank of the matroid. If all dependent sets C of M have size $|C| \geq r(M)$, then M is a paving matroid.

2.5.2   EXAMPLE

Let M be a matroid with ground set S and set of bases B given by

$$S = \{a, b, c\}$$

$$B = \{\{a, b\}, \{a, c\}, \{b, c\}\}$$

Show that M is a paving matroid.

First, we need to determine the rank of M. The rank is the cardinality of any of the bases. We see from the definition that every basis has cardinality 2, so $r(M) = 2$.

We need to identify the dependent sets and check their cardinality. Now, every basis is an independent set, and every subset of a basis is also an independent set. In fact, these are all of the independent sets of M.

In this case, the only subset of S that is not an independent set is S itself. The cardinality of S is $|S| = 3 > r(M)$, so M is a paving matroid because every dependent set C satisfies $|C| \geq r(M)$.

2.5.3   EXAMPLE

Let M be a matroid with ground set S and set of bases B given by

$$S = \{a, b, c, d\}$$

$$B = \{\{a, b\}, \{a, c\}, \{b, c\}\}$$

Show that M is a not paving matroid.

The set of basis is exactly the same as the previous example, so the rank of M is $r(M) = 2$.

In fact, the only difference between this and the previous example is the ground set. This example has an additional element d in its ground set.

However, d does not appear in any of the bases of M. Thus, $\{d\}$ must be a dependent set of M. But $|\{d\}| = 1 < r(M)$.

We have identified a dependent set whose cardinality is less than the rank of the matroid. Thus, M is not a paving matroid.

2.5.4 EXAMPLE

Let M be a matroid with ground set S and set of bases B given by

$$S = \{a, b, c\}$$

$$B = \{\{a, b\}, \{a, c\}\}$$

Show that M is a paving matroid.

We see that the rank of M is $r(M) = 2$. The only dependent sets of M are $\{a, b, c\}$ and $\{b, c\}$. The cardinality of these sets is

$$|\{a, b, c\}| = 3 \quad |\{b, c\}| = 2$$

In this case, we have a dependent set whose cardinality is equal to the rank of the matroid. This is allowed because the dependent sets of a paving matroid must have cardinality greater than or equal to the cardinality of the matroid.

Thus, M is a paving matroid because the cardinality of the dependent sets all satisfy $|C| \geq r(M)$.

2.5.5 THEOREM

Let M be a uniform matroid. Then M is a paving matroid.

Proof. Recall that a uniform matroid is a matroid whose set of bases is all m element subsets of the ground set. However, every k element set must be a subset of one of these basis sets if $k \leq m$. In this case, the independent sets are all subsets of the ground set whose cardinality is less than or equal to m.

The dependent sets are all subsets of the ground set whose cardinality is greater than m. The rank of M is the cardinality of any of the basis sets so $r(M) = m$. But every dependent set has cardinality greater than m. Therefore, M is a paving matroid. ∎

2.5.6   EXAMPLE

Let M be a matroid with ground set S and set of bases B given by

$$S = \{a, b, c, d\}$$

$$B = \{\{a, b\}, \{a, c\}, \{b, c\}, \{b, d\}, \{c, d\}\}$$

This matroid is $U_{2,4}$. We see that every two-element subset of S is a basis of M. Since the basis sets have cardinality 2, the rank of the matroid is $r(M) = 2$. Furthermore, any set with cardinality ≤ 2 is an independent set because these sets are all subsets of some basis set.

With this, we see that all dependent sets must have cardinality > 2. Because all dependent sets have cardinality $\geq r(M) = 2$, M is a paving matroid.

2.6   Partition Matroids

A partition matroid is a matroid whose ground set can be divided into a series of disjoint sets where the independent sets are formed from the Cartesian product of the subsets of disjoint sets. Partition matroids are useful in modeling bipartite matching problems.

2.6.1   DEFINITION: PARTITION MATROID

Let $S_1, S_2, \ldots, S_n$ be a series of disjoint sets. Let $d_1, d_2, \ldots, d_n$ be a series of positive integers where $d_i \leq |S_i|$. There is a partition matroid M with ground set $S = \bigcup_{i=1}^{n} S_i$ and independent set

$$\mathfrak{I} = \{I \subseteq S \mid |I \cap S_i| \leq d_i, i = 1 \ldots n\}$$

We form a matroid from a series of n disjoint sets S_i. by specifying a maximum amount of elements d_i we are allowed to choose from each set. The ground set of the matroid is the union of the disjoint sets. The independent sets are the sets where $|I \cap S_i| \leq d_i$ for each value of i.

2.6.2   THEOREM

Let M be a partition matroid. The maximal independent sets of M have cardinality $\sum_{i=1}^{n} d_i$.

Proof. Let B be a maximal independent set of M. Since B is an independent set, we know that $|I \cap S_i| \leq d_i$ for every i. But since B is a maximal independent set, the equality must hold so $|I \cap S_i| = d_i$.

However, because the S_i's are disjoint, the sets $I \cap S_i$ are also disjoint. Thus, we have d_i distinct elements in the basis for every $I \cap S_i$. Furthermore, since $S = \bigcup_{i=1}^{n} S_i$ and $I \subseteq S$, this must be all of the elements in the set.

Therefore, we can find the cardinality of the maximal independent set by adding up the values of the d_i's,

$$|B| = \sum_{i=1}^{n} d_i \ \blacksquare$$

2.6.3   THEOREM

A partition matroid is in fact a matroid.

Proof. We show that the maximal independent sets from a partition matroid satisfy **BM-1** and **BM-2** of a basis matroid. Let the set B be the set of sets in $\mathfrak{I}$ with cardinality $\sum_{i=1}^{n} d_i$.

BM-1 $B \neq \emptyset$

Because $0 < d_i \leq |S_i|$ for every i, there must be at least one set $I \subseteq S$ satisfying $\{I \subseteq S \mid |I \cap S_i| \leq d_i, i = 1 \ldots n\}$. Thus, the set of all bases B cannot be empty.

BM-2 For every distinct $B_1, B_2 \in B$, and every element $b \in B_1 - B_2$, there exists a element $\bar{b} \in B_2 - B_1$ such that $(B_1 - b) \cup \bar{b} \in B$

We have a set of bases for the partition matroid that all have cardinality $\sum_{i=1}^{n} d_i$. Let B_1, B_2 be distinct bases. Because B_1 and B_2 are distinct and have the same cardinality, $|B_1 - B_2| > 0$ and $|B_2 - B_1| > 0$.

Let $b \in B_1 - B_2$ be an element of $B_1 - B_2$. b must belong to one of the disjoint sets S_i. Designate this set as S_b, and let d_b be the value of d_i for the set S_b. Now, both $B_1 \cap S_b$ and $B_2 \cap S_b$ have the same number of elements d_b because both B_1 and B_2 are maximal sets. However, as shown in Figure 13, if $|(B_1 - B_2) \cap S_b| > 0$ then $|(B_2 - B_1) \cap S_b| > 0$.

We see that if there is an element $b \in B_1 - B_2$ where $b \in S_b$, then there must be some element $\underline{b} \in B_1 - B_2$ where $\underline{b} \in S_b$. So, $b \in B_1$ but $b \notin B_2$, $\underline{b} \in B_2$ but $\underline{b} \notin B_1$, and $b, \underline{b} \in S_b$. This situation is depicted in Figure 13.

Now, form $B_e = (B_1 - b) \cup \underline{b}$. The only difference between B_1 and B_e is that we have removed b and inserted $\underline{b}$. But because $\underline{b}$ was not part of B_1 originally, we know $|B_e| = |B_1|$.

For every set S_i, $|B_e \cap S_i| = d_i$. This is certainly true for every $S_i \neq S_b$ because $B_e \cap S_i = B_1 \cap S_i$, and we know that $|B_1 \cap S_i| = d_i$ because B_1 is a maximal independent set. In addition, we know $|B_e \cap S_b| = |B_1 \cap S_b|$ because when we constructed B_e, we removed b from B_1 and then inserted a new element $\underline{b}$, and $b, \underline{b} \in S_b$.

B$_1$

| a$_1$ | a$_2$ | ... | a$_{d1}$ | ... | b$_1$ | b$_2$ | ... | b | ... | ... |

B$_2$

| a̲$_1$ | a̲$_2$ | ... | a̲$_{d1}$ | ... | b̲$_1$ | b̲$_2$ | ... | b̲ | ... | ... |

B$_e$=(B$_1$-b)U b̲

| a$_1$ | a$_2$ | ... | a$_{d1}$ | ... | b$_1$ | b$_2$ | ... | b̲ | ... | ... |

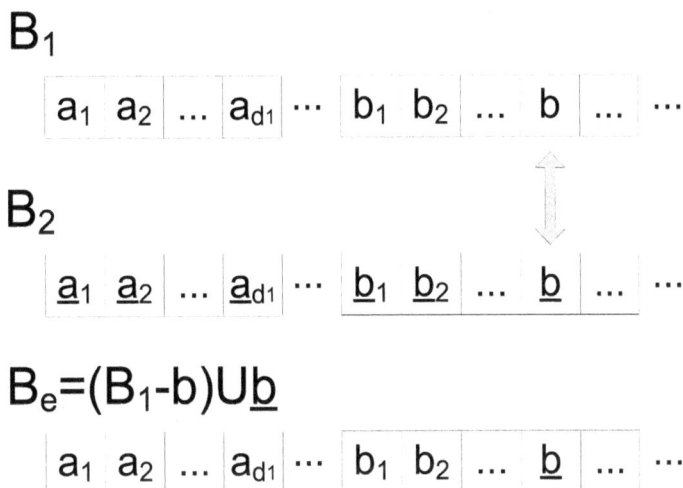

Figure 13: Construction of a new basis from a pair of bases.

However, since $|B_e \cap S_i| = d_i$ for every i, then B_e must be a maximal independent set. This means that $B_e \in B$. Thus, starting from distinct elements $B_1, B_2 \in B$, for every element $b \in B_1 - B_2$, there exists a element $\bar{b} \in B_2 - B_1$ such that $(B_1 - b) \cup \bar{b} \in B$.■

2.6.4 EXAMPLE

Find the ground set and set of bases for a partition matroid formed from the disjoint sets

$$S_1 = \{a, b\} \quad S_2 = \{c, d\}$$

where $d_1 = 1$ and $d_2 = 2$.

The ground set of the partition matroid is the union of the disjoint sets,

$$S = \bigcup_{i=1}^{n} S_i = S_1 \cup S_2 = \{a, b, c, d\}.$$

The set of bases is all independent sets with cardinality $|I| = \sum_{i=1}^{n} d_i = 3$. We can form these sets as the Cartesian product of all sets of S_1 with cardinality d_1 and all sets of S_2 with cardinality d_2.

$$\{\{a\}, \{b\}\} \times \{\{c, d\}\} = \{\{a, c, d\}, \{b, c, d\}\}.$$

2.6.5 PROPOSITION

Let M be a partition matroid formed from the disjoint sets $S_1, S_2, \dots, S_n$ with $d_1, d_2, \dots, d_n$ where $0 < d_i \leq |S_i|$. The number of basis sets of M is

$$|B| = \prod_{i=1}^{n} \binom{|S_i|}{d_i}.$$

Proof. Each basis of M is formed as a product of distinct elements from each of the disjoint sets. Every distinct product produces a distinct basis.

We can compute the total number of products by multiplying together the total number of distinct elements sets of cardinality d_i for each of the S_i's. For a given S_i, the total number of distinct sets of d_i elements is given by the combinatorial $\binom{|S_i|}{d_i}$. The total number of bases is computed by multiplying all of these factors together, which leads to the desired result. ∎

2.7   Representable Matroid

A representable matroid is a vector matroid that can be expressed in terms of some field. A matroid that can be expressed over $\mathbb{Z}_2$ is called a binary matroid, while a matroid expressed over $\mathbb{Z}_3$ is called a ternary matroid.

2.7.1 DEFINITION: REPRESENTABLE MATROID

> Let M be a vector matroid. If there exists a matrix representation of M over some field $\mathcal{F}$, then M is a representable over $\mathcal{F}$.

2.7.2 EXAMPLE

Let M be a matroid with ground set S and set of bases B

$$S = \{a, b, c\}$$

$$B = \{\{a, b\}, \{a, c\}, \{b, c\}\}$$

Determine if M is representable over $\mathbb{Z}_2$.

First, we need to recall that the field $\mathbb{Z}_2$ has two elements. We may represent these elements in number of different ways, but let's choose $\{0,1\}$. The algebra of $\mathbb{Z}_2$ under addition is

$$0 + 0 = 0 \quad 0 + 1 = 1$$
$$1 + 0 = 1 \quad 1 + 1 = 0$$

We treat subtraction as the inverse of addition and have

$$0 - 0 = 0 \quad 0 - 1 = 1$$
$$1 - 0 = 1 \quad 1 - 1 = 0$$

We desire a matrix representation of M with three vectors using only the elements $\{0,1\}$, the elements of $\mathbb{Z}_2$. If B is the basis then any two vectors must be linearly independent.

Choose,

$$v_a = \begin{pmatrix} 1 \\ 0 \end{pmatrix} \quad v_b = \begin{pmatrix} 0 \\ 1 \end{pmatrix} \quad v_c = \begin{pmatrix} 1 \\ 1 \end{pmatrix}$$

We can see that any two of these vectors are linearly independent, while all three together are dependent. The matrix form for the matroid is

$$\begin{pmatrix} 1 & 0 & 1 \\ 0 & 1 & 1 \end{pmatrix}$$

We have constructed a matrix representation of M using only $\mathbb{Z}_2$ so M is representable over $\mathbb{Z}_2$.

2.7.3 DEFINITION: BINARY MATROID

Let M be a vector matroid. If M is representable over the field $\mathbb{Z}_2$, then M is a binary matroid.

2.7.4 DEFINITION: TERNARY MATROID

Let M be a vector matroid. If M is representable over the field $\mathbb{Z}_3$, then M is a ternary matroid.

2.7.5   EXAMPLE

The matroid in example 2.7.2 is both binary and ternary. In the example, we constructed a representation of M over the field $\mathbb{Z}_2$, so M must be binary.

Furthermore, the matroid is representable over $\mathbb{Z}_3$. Choose vectors

$$v_a = \begin{pmatrix} 1 \\ 0 \end{pmatrix} \quad v_b = \begin{pmatrix} 0 \\ 1 \end{pmatrix} \quad v_c = \begin{pmatrix} 1 \\ -1 \end{pmatrix}$$

Again, any pair of vectors is independent while the triple is dependent. The matroid can be represented by a matrix over $\mathbb{Z}_3$ as

$$\begin{pmatrix} 1 & 0 & 1 \\ 0 & 1 & -1 \end{pmatrix}$$

Thus, this matroid is both binary and ternary.

2.7.6   EXAMPLE

Let M be a matroid with ground set S and set of bases B

$$S = \{a, b, c, d\}$$

$$B = \{\{a, b\}, \{a, c\}, \{a, d\}, \{b, c\}, \{b, d\}, \{c, d\}\}$$

Determine if M is binary and/or ternary.

We need to find four distinct vectors over $\mathbb{Z}_2$ such that every pair is independent but every triple is dependent. We can force every triple to be dependent by choosing to use two-dimensional vectors.

However, there are only four two-dimensional vectors over $\mathbb{Z}_2$:

$$v_a = \begin{pmatrix} 1 \\ 0 \end{pmatrix} \quad v_b = \begin{pmatrix} 0 \\ 1 \end{pmatrix} \quad v_c = \begin{pmatrix} 1 \\ 1 \end{pmatrix} \quad v_d = \begin{pmatrix} 0 \\ 0 \end{pmatrix}$$

v_d is not independent of all other vectors. Thus, every pair is not independent, so this representation over $\mathbb{Z}_2$ is invalid. But if we use a three-dimensional vector, we either arrive as an equivalent set of vectors, or we have triples that are independent. The same holds for higher dimensional representations. Thus, the matroid is not representable over $\mathbb{Z}_2$.

The situation is different for ternary vectors. With a third value, we can use the vectors

$$v_a = \begin{pmatrix} 1 \\ 0 \end{pmatrix} \quad v_b = \begin{pmatrix} 0 \\ 1 \end{pmatrix} \quad v_c = \begin{pmatrix} 1 \\ 1 \end{pmatrix} \quad v_d = \begin{pmatrix} 1 \\ -1 \end{pmatrix}$$

We see that every pair of vectors is independent. Furthermore, any triple is dependent. This matroid is representable over $\mathbb{Z}_3$, so it is ternary.

2.8   Transversal Matroids

A transversal matroid is class of matroids formed from the set of partial transversals on a set. This class of matroids is useful in analyzing combinatorial matching problems.

2.8.1   DEFINITION: TRANSVERSAL MATROID

> Let S be a set, and C be a sequence of subsets where $C_i \subseteq S$ (not necessarily distinct). Let P be the set of all partial transversals of C. Then (S, P) forms a matroid.

In a transversal matroid, the independent set of the matroid is taken as the set of all partial transversals of C.

Proof. We show that the axioms of a set matroid are satisfied if we take $T = P$ as the independent set.

SM-1 $\emptyset \in T$

T is the set of all partial transversals of C. But $\emptyset$ is a valid partial transversal of any set. So $\emptyset \in T$.

SM-2   If $t \in T$ and $\bar{t} \subseteq t$, then $\bar{t} \in T$

If $p \in P$ is a partial transversal of C, then any subset of p is also a partial transversal. If $\bar{p} \subseteq p$, then $\bar{p}$ is also a partial transversal, so $\bar{p} \in P$. Since $T = P$, **SM-2** is satisfied.

SM-3   Let $t_1, t_2 \in T$ where $|t_1| < |t_2|$, then $\exists\, d \in t_2 - t_1$ such that $t_1 \cup d \in T$

This requirement is more difficult. We need to show that partial transversals satisfy **SM-3**. The sequence C is a sequence of sets where a given partial transversal may choose exactly one member from the set (or it may decide to not choose any members of the set). For the moment we will call the individual members of C 'switches'.

Examine two partial transversals A and B where $|A| < |B|$. Each partial transversal has some set of switches turned to a specific value. We put each switch into one of four Types: I) switch is used in A but not B; II) switch is used in B but not A; III) switch us used in both A and B; IV) switch is not used in either A or B.

We further divide the Type III switches into two subcategories: IIIs) both A and B have the same value for the switch (s for same); IIId) A and B have different values for the switch (d for different).

Now, look at the elements in $d \in B - A$. If any of these elements are present in any switch of Type II or Type IV, then we simply take all of the A switches initially set, turn one Type II/IV switch to any element in $d \in$B-A, and we have constructed a partial transversal $A \cup d$, satisfying **SM-3**. This process works because the partial transversal A does not use any of the Type II/IV switches, so we can turn one of these on without disrupting the original elements in A.

What if there are no Type II/IV switches containing an element $d \in B - A$? Then the B sequence must use a Type IIId switch. This must be true because the B sequence does not use any Type I switches, and it can't be a Type IIIs switch because d must be in B but not A, and all of the elements using Type IIIs switches are in both B and A.

We can form ordered pairs (a, b) from the Type IIId switches, where $a \in A$, $b \in B$, and the ordered pair represents the different choices made by A versus B for the same switch. From above, we know that there must be at least $|B - A|$ Type IIId switches (otherwise **SM-3** is immediately satisfied as shown earlier). Also, every element $d \in B - A$ is present on the b-side of an ordered pair of some Type IIId switch.

Next, construct a sequence of ordered pairs of Type IIId in the following manner:

1. Choose any $d \in B - A$
2. Find the Type IIId element with d on the b-side
3. Identify the element on the a-side; call it x
4. Look for a Type IIId element with x on the b-side
    a. If there is such an element, go back to step 3
    b. If there is not such an element, go to step 5
5. We have arrived at some x where there is no Type IIId switch with x on the b-side. However, there is a Type IIId switch with x on the a-side.

We have a sequence $(a_1, d), (a_2, a_1), (a_3, a_2) \ldots (x, a_n)$. This sequence must terminate after a finite number of steps. First, the sequence produced by 4a cannot form a cycle because each element can only appear on a given side only once. Second, the sequence must eventually terminate because there are only a finite number of elements in S.

The sequence is unique and can be traversed in either direction. The entire sequence may be discovered from any a-element or any b-element. We can also uniquely identify the sequence by specifying the beginning and ending elements as (x, d).

Since x appears on the a-side of the ordered pair, $x \in A$. Now, divide the set A into two disjoint subsets: $A - B$ and $A \cap B$. We must have either $x \in A - B$ or $x \in A \cap B$.

Suppose $x \in A \cap B$, then x is also in B. But since there is no (y, x) among the Type IIId switches, then x must be on one of the Type II switches. But since the Type II switches are not used by A, we can turn this switch on, and add it to A. Now change every switch in the sequence into its corresponding b-value. Combining this with the other A switches, **SM-3** is satisfied.

Figure 14 demonstrates this technique. If the sequence eventually terminates with an element $x \in A \cap B$, then there is some switch in B set to the value x that is not a switch used in A. If we set this switch to x, then choose all of the common switches in the sequence to have the value from B instead of A, we get the A sequence plus the additional element d. Thus, we have identified a partial transversal $A \cup d$, satisfying **SM-3**.

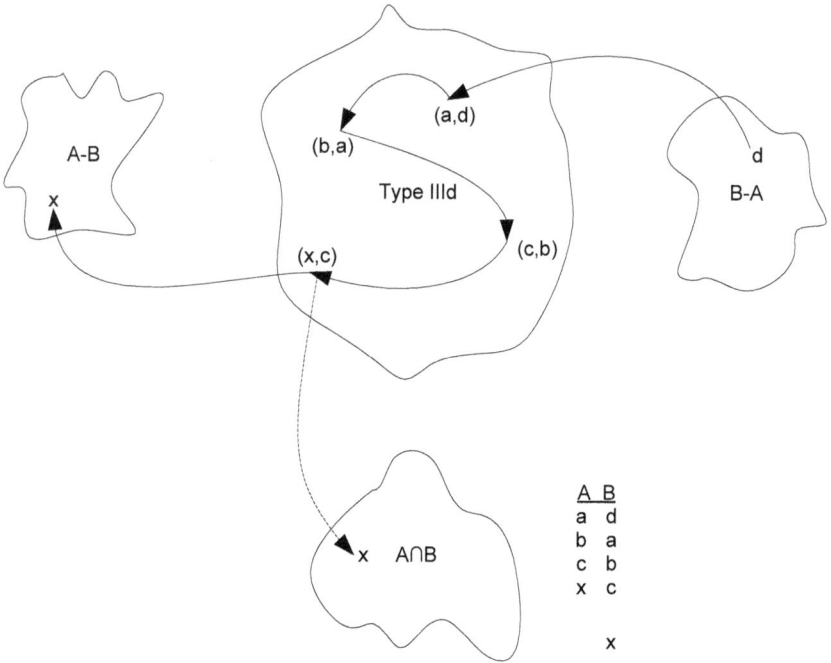

Figure 14: Transversal.

If $x \in A - B$, create the unique ordered pair (x, d) indicating that the initial choice of $d \in B - A$ terminated with $x \in A - B$. Next, choose another $d \in B - A$ and repeat from step 1. If every d terminates with an $x \in A - B$, then we will have a unique $x \in A - B$ associated with a unique $d \in B - A$. In this case we must have $|A - B| \geq |B - A|$, because for every $d \in B - A$ we have identified a distinct $x \in A - B$. However, since $|A| = |A - B| + |A \cap B|$ and $|B| = |B - A| + |A \cap B|$, the initial assertion that $|A| < |B|$ is equivalent to $|A - B| < |B - A|$. This contradicts the conclusion that $|A - B| \geq |B - A|$, so **SM-3** must be satisfied by some element $d \in B - A$. ∎

2.8.2   EXAMPLE

Let $S = \{a, b, c, d\}$. Let C be the sequence of subsets

$$C_1 = \{a, c, d\} \quad C_2 = \{b, c\} \quad C_3 = \{b, c\}$$

Find the transversal matroid from C.

First, we need to find all partial transversals of C. Below we list the partial transversals of C using the notation $[x, y, z]$ where x is the element choice from C_1, y from C_2, and z from C_3. If we do not select an element from a particular subset, we denote this as $\emptyset$.

$$[a, \emptyset, \emptyset] \quad [a, b, \emptyset] \quad [a, c, \emptyset] \quad [a, \emptyset, b] \quad [a, \emptyset, c] \quad [a, b, c] \quad [a, c, b]$$

$$[c, \emptyset, \emptyset] \quad [c, b, \emptyset] \quad [c, \emptyset, b]$$

$$[d, \emptyset, \emptyset] \quad [d, b, \emptyset] \quad [d, c, \emptyset] \quad [d, \emptyset, b] \quad [d, \emptyset, c] \quad [d, b, c] \quad [d, c, b]$$

$$[\emptyset, \emptyset, \emptyset]$$

As a collection of sets, the partial transversals are:

$$\emptyset \quad \{a\} \quad \{b\} \quad \{c\} \quad \{d\} \quad \{a, b\} \quad \{a, c\} \quad \{b, c\} \quad \{b, d\} \quad \{c, d\}$$

$$\{a, b, c\} \quad \{b, c, d\}$$

All of these sets are subsets of $\{a, b, c\}$ and $\{b, c, d\}$. We have a transversal matroid M with ground set S and set of bases B

$$S = \{a, b, c, d\}$$
$$B = \{\{a, b, c\}, \{b, c, d\}\}$$

Chapter 3: Matroid Properties

We explore several matroid properties including matroid rank and closure. In addition, we define the basis of a matroid. The basis is the matroid analog of the linear algebra concept of basis vectors in a vector space. These concepts are very useful in identifying matroids as well as proving several matroid theorems.

3.1 Basis

The concept of a basis for a matroid is an extension of the basis concept in vector spaces. In fact, there are several similarities between a matroid basis and a basis in a vector space. This section will explore some of these concepts.

3.1.1 DEFINITION: BASIS I

Let M be a matroid (S, T). A basis of M is a maximally independent set of T.

The set B of all possible bases is the set of all maximally independent sets of T. Alternatively, given B, T may be constructed as the set of all subsets of elements in B. This provides yet another method for constructing a matroid.

3.1.2 EXAMPLE

Find every basis for the vector matroid from the 2×3 matrix

$$\begin{bmatrix} 1 & 0 & 1 \\ 0 & 1 & 1 \end{bmatrix}.$$

First, find the independent sets of the matroid. Label the columns a, b and c respectively; $\Lambda = \{a, b, c\}$ with corresponding vectors $v_a = \begin{pmatrix} 1 \\ 0 \end{pmatrix}$, $v_b = \begin{pmatrix} 0 \\ 1 \end{pmatrix}$ and $v_c = \begin{pmatrix} 1 \\ 1 \end{pmatrix}$. The set of all sets of linearly independent column headers is $I = \{\emptyset, \{a\}, \{b\}, \{c\}, \{a, b\}, \{a, c\}, \{b, c\}\}$.

The sets $\{a, b\}, \{a, c\}$, and $\{b, c\}$ are all maximally independent sets. Each of these sets is a basis for M.

3.1.3   EXAMPLE

Find every basis for the vector matroid from the 3×5 matrix

$$\begin{bmatrix} 1 & 1 & 0 & 0 & 2 \\ 0 & 1 & 0 & 1 & 0 \\ 0 & 1 & 0 & 1 & 0 \end{bmatrix}.$$

Label the columns a, b, c, d and e respectively; $\Lambda = \{a, b, c, d, e\}$ with corresponding vectors $v_a = \begin{pmatrix} 1 \\ 0 \\ 0 \end{pmatrix}$, $v_b = \begin{pmatrix} 1 \\ 1 \\ 1 \end{pmatrix}$, $v_c = \begin{pmatrix} 0 \\ 0 \\ 0 \end{pmatrix}$, $v_d = \begin{pmatrix} 0 \\ 1 \\ 1 \end{pmatrix}$,

and $v_e = \begin{pmatrix} 2 \\ 0 \\ 0 \end{pmatrix}$. The set of all sets of linearly independent column headers is

$$I = \left\{ \begin{array}{c} \emptyset, \{a\}, \{b\}, \{d\}, \{e\}, \{a, b\}, \{a, d\}, \{b, d\}, \\ \{b, e\}, \{d, e\}, \{a, b, d\}, \{b, d, e\} \end{array} \right\}.$$

The sets $\{a, b, d\}$, and $\{b, d, e\}$ are both maximally independent sets and both provide a basis for M.

3.1.4   EXAMPLE

Find every basis for the graphic matroid from Example 2.3.3.

The cycle set is $\Gamma = \{\{a, b, c\}\}$, and the power set is

$$P = \{\emptyset, \{a\}, \{b\}, \{c\}, \{a, b\}, \{a, c\}, \{b, c\}, \{a, b, c\}\}$$

The independent sets are the elements of $I = P - \Gamma$ or

$$I = \{\emptyset, \{a\}, \{b\}, \{c\}, \{a, b\}, \{a, c\}, \{b, c\}\}.$$

The maximally independent sets are $\{a, b\}, \{a, c\}$, and $\{b, c\}$, which forms the bases for the graphic matroid M.

3.1.5   EXAMPLE

Find every basis for the graphic matroid from Figure 15.

The cycle set is $\Gamma = \{\{a,b,c\}, \{d,c,e\}, \{a,b,d,e\}\}$. The maximally independent sets are:

$$\{a,b,d\}, \{a,b,e\}, \{a,b,d\}, \{a,b,e\}, \{a,c,d\},$$

$$\{a,c,e\}, \{a,d,e\}, \{b,c,d\}, \{b,c,e\}, \{b,d,e\}$$

3.1.6 THEOREM

If B_1 and B_2 are both bases for a matroid M, then $|B_1| = |B_2|$.

Proof. If B_1 is a basis for M, then B_1 is a maximally independent set. Suppose $|B_1| < |B_2|$. By property **SM-3**, there is a $d \in B_2 - B_1$ such that $B_1 \cup d \in T$. Let $B_3 = B_1 \cup d$. Since $B_3 \in T$, B_3 is an independent set. However, $B_1 \subset B_3$, so B_1 is a proper subset of another independent set B_3. This contradicts the assertion that B_1 is basis of M. ∎

3.1.7 THEOREM

A matroid has at least one basis.

Proof. A basis of a matroid is a maximally independent set. However, by **SM-1**, the independent set of the matroid has at least one member. Since there is at least one independent set, there is at least one maximally independent set, so there must be at least one basis. ∎

3.1.8 THEOREM

Let B be the set of bases of a matroid M, and let $B_1, B_2 \in B$ be distinct bases. Let $b \in B_1 - B_2$. There exists a $\bar{b} \in B_2 - B_1$ such that $(B_1 - b) \cup \bar{b} \in B$.

Proof. The members of B are independent sets. By property **SM-2**, every subset of an independent set is also an independent set. So $B_1 - b$ is an independent set.

Furthermore, since B_1 and B_2 are both bases, $|B_1| = |B_2|$ by theorem 3.1.5. Then it must be that $|B_1 - b| < |B_1|$.

Using **SM-3**, there must be some element $\bar{b} \in B_2 - B_1$ such that $(B_1 - b) \cup \bar{b}$ is an independent set. This independent set must be contained in some maximally independent set B_{max}.

But $\left|(B_1 - b) \cup \bar{b}\right| = |B_1|$, and all bases must have the same dimension. There cannot be a larger independent set B_{max} with $(B_1 - b) \cup \bar{b} \subset B_{max}$ because then $|B_{max}| > |B_1|$. So it must be that $B_{max} = (B_1 - b) \cup \bar{b}$, making $(B_1 - b) \cup \bar{b}$ a maximally independent set.

Putting everything together, if $B_1, B_2 \in B$ and $b \in B_1 - B_2$, then there is an element $\bar{b} \in B_2 - B_1$ with the property that $(B_1 - b) \cup \bar{b} \in B$.■

3.1.9   DEFINITION: BASIS II

Let M be a matroid (S, T), and let B be a collection of subsets of S satisfying:

BM-1 $B \neq \emptyset$

BM-2 For every distinct $B_1, B_2 \in B$, and every element $b \in B_1 - B_2$, there exists an element $\bar{b} \in B_2 - B_1$ such that $(B_1 - b) \cup \bar{b} \in B$

Then B is the set of bases of M.

3.1.10   EXAMPLE

Let $B = \{\{a, b\}, \{a, c\}, \{b, c\}\}$ be the set of bases for a matroid M. Show that these sets satisfy the basis relations, and find the independent sets of M.

BM-1

Because there is at least one element in B, $B \neq \emptyset$.

BM-2

Examine each pair of bases in B and show that **BM-2** is satisfied for each.

Case 1: $B_1 = \{a, b\}$, $B_2 = \{a, c\}$

$$D = B_1 - B_2 = \{b\} \quad \bar{D} = B_2 - B_1 = \{c\}$$

If we remove b from B_1 and replace it with c, we have $\{a, c\} \in B$.

Case 2: $B_1 = \{a, c\}$, $B_2 = \{a, b\}$

$$D = B_1 - B_2 = \{c\} \quad \bar{D} = B_2 - B_1 = \{b\}$$

Removing c from B_1 and replacing it with b, we have $\{a, b\} \in B$.

Case 3: $B_1 = \{a, b\}$, $B_2 = \{b, c\}$

$$D = B_1 - B_2 = \{a\} \quad \bar{D} = B_2 - B_1 = \{c\}$$

We have $\{b, c\} \in B$.

Case 4: $B_1 = \{b, c\}$, $B_2 = \{a, b\}$

$$D = B_1 - B_2 = \{c\} \quad \bar{D} = B_2 - B_1 = \{a\}$$

We have $\{a, b\} \in B$.

Case 5: $B_1 = \{a, c\}$, $B_2 = \{b, c\}$

$$D = B_1 - B_2 = \{a\} \quad \bar{D} = B_2 - B_1 = \{b\}$$

We have $\{b, c\} \in B$.

Case 6: $B_1 = \{b, c\}$, $B_2 = \{a, c\}$

$$D = B_1 - B_2 = \{b\} \quad \bar{D} = B_2 - B_1 = \{a\}$$

We have $\{a, c\} \in B$.

In each case, **BM-2** is satisfied, so B satisfies the basis relations.

The independent sets of M are the union of all subsets of B. The subsets of the bases in B are:

$$\{a, b\} \supseteq \emptyset, \{a\}, \{b\}, \{a, b\}$$

$$\{a, c\} \supseteq \emptyset, \{a\}, \{c\}, \{a, c\}$$

$$\{b,c\} \supseteq \emptyset, \{b\}, \{c\}, \{b,c\}$$

So the independent sets are $T = \{\emptyset, \{a\}, \{b\}, \{c\}, \{a,b\}, \{a,c\}, \{b,c\}\}$.

3.1.11 DEFINITION: BASIS MATROID

> Let S be a set and let B be a collection of subsets of S where B satisfies
>
> **BM-1** $B \neq \emptyset$
>
> **BM-2** For every distinct $B_1, B_2 \in B$, and every element $b \in B_1 - B_2$, there exists an element $\bar{b} \in B_2 - B_1$ such that $(B_1 - b) \cup \bar{b} \in B$
>
> Let $\mathfrak{I}$ be the collection of all subsets of the elements of B. Then the ordered pair $(S, \mathfrak{I})$ is a matroid.

Proof. We proceed to show that the above definition satisfies **SM-1**, **SM-2**, and **SM-3**.

SM-1 $\emptyset \in \mathfrak{I}$

From the definition, $\mathfrak{I}$ contains all of the subsets of the elements of B. But form **BM-1**, we see that B is not empty. There must be some set $B_1 \in B$ in B, and $\mathfrak{I}$ contains every subset of B_1. Because $\emptyset \subseteq B_1$, $\emptyset \in \mathfrak{I}$.

SM-2   If $t \in \mathfrak{I}$ and $\bar{t} \subseteq t$, then $\bar{t} \in \mathfrak{I}$

If $t \in \mathfrak{I}$, then t must be a subset of some element $B_1 \in B$. If $\bar{t} \subseteq t$, and $t \subseteq B_1$, then $\bar{t} \subseteq B_1$. But by definition, every subset of B_1 is also a member of $\mathfrak{I}$. Thus, since $\bar{t} \subseteq B_1$ then $\bar{t} \in \mathfrak{I}$.

SM-3   Let $t_1, t_2 \in \mathfrak{I}$ where $|t_1| < |t_2|$, then $\exists d \in t_2 - t_1$ such that $t_1 \cup d \in \mathfrak{I}$

Assume that **SM-3** does not hold for some $t_1, t_2 \in \mathfrak{I}$. There must be elements $B_1, B_2 \in B$ such that $t_1 \subseteq B_1$ and $t_2 \subseteq B_2$. There may be many possible choices for B_2. Choose B_2 such that $|B_2 - I_2 \cup B_1|$ is minimal.

Let $d \in t_2 - t_1$. If **SM-3** does not hold, then $d \notin B_1$ because then $t_1 \cup d \subseteq B_1$ and every subset of $B_1 \in \Im$ by definition. In this case, $t_2 - t_1 = t_2 - B_1$.

Let $x \in B_2 - t_2 \cup B_1$. x must also be an element of $B_2 - B_1$. From **BM-2** there must be $y \in B_1 - B_2$ such that $(B_2 - x) \cup y \in B$. Set $B_3 = (B_2 - x) \cup y$. We construct B_3 by removing x from B_2 and adding y. Since, $t_2 \subseteq B_2$, and since $x \notin t_2$, $t_2 \subseteq B_3$. Now, B_3 has one additional element, y, in common with $t_2 \cup B_1$ than does B_2. But then $|B_3 - t_2 \cup B_1| < |B_2 - t_2 \cup B_1|$, which contradicts the assumption that $|B_2 - t_2 \cup B_1|$ is minimal. Thus, it must be the case that $B_2 - t_2 \cup B_1 = \emptyset$.

Now, if $B_2 - t_2 \cup B_1 = \emptyset$ then $B_2 - B_1 = t_2 - B_1$. Essentially, since $B_2 = t_2 \cup B_1$, $B_2 - B_1 = t_2 \cup B_1 - B_1$, but $t_2 \cup B_1 - B_1 = t_2 - B_1$. Combining this with the fact that $t_2 - t_1 = t_2 - B_1$, we see that $B_2 - B_1 = t_2 - t_1$.

Next, examine $B_1 - t_1 \cup B_2$. Let $x \in B_1 - t_1 \cup B_2$. By the same argument as above, we have $x \in B_1 - B_2$ and from **BM-2**, there must be a $y \in B_2 - B_1$ such that $(B_1 - x) \cup y \in B$. Now, since $t_1 \subseteq B_1$ and $x \notin t_1$, we know $t_1 \cup y \subseteq (B_1 - x) \cup y$. But form BM-2, $(B_1 - x) \cup y \in B$ which means that $(B_1 - x) \cup y \in \Im$. But this contradicts the assumption that **SM-3** does not hold. It can only be the case that $B_1 - t_1 \cup B_2 = \emptyset$.

Again, similar to the reasoning above, if $B_1 - t_1 \cup B_2 = \emptyset$ then $B_1 - B_2 = t_1 - B_2$. But since $t_2 \subseteq B_2$, then $t_1 - B_2 \subseteq t_1 - t_2$. Consequently $|t_1 - B_2| \leq |t_1 - t_2|$ which means that $|B_1 - B_2| \leq |t_1 - t_2|$.

Based on **BM-2**, we know that the elements of B are all equicardinal. To show this, assume that there exist two distinct members $C, D \in B$ such that $|C| < |D|$. Furthermore, if there are multiple such pairs of C and D, choose C, D such that $|D - C|$ is minimal. Let $x \in D - C$. From **BM-2** there must exist $y \in C - D$ such that $(D - x) \cup y \in B$. Set $E = (D - x) \cup y$. Now, we know $|E| = |D|$ and since E has an additional element in common with C, $|E - C| < |D - C|$. But this contradicts the choice of $|D - C|$ as minimal. Thus, the initial assumption that $|C| < |D|$ is invalid.

But if all elements of B are equicardinal, then $|B_1 - B_2| = |B_2 - B_1|$. Thus, $|B_1 - B_2| = |B_2 - B_1| \le |t_1 - t_2|$, while $|B_2 - B_1| = |t_2 - t_1|$, so $|t_2 - t_1| \le |t_1 - t_2|$. However, this means that $|t\_2| \le |t\_1|$, which contradicts our initial assumption that $|t_1| < |t_2|$. With this final contradiction, we find that the assumption that **SM-3** does not hold is invalid. ∎

3.1.12 DEFINITION: ISTHMUS

> Let M be a matroid (S, T) with set of bases B. If $e \in S$ is in every basis, i.e. $e \in b \ \forall \ b \in B$, then e is called an isthmus.

An isthmus is an element that is in every basis of M. Isthmuses will be important in the study of rank as related to matroid invariants.

3.1.13 EXAMPLE

Find all elements from Example 3.1.3 that are isthmuses.

We found the set of bases in Example 3.1.3 is $B = \{\{a, b, d\}, \{b, d, e\}\}$. There are two elements present in every basis, b and d. Both of these are isthmuses of M.

3.1.14 PROPOSITION

Let M be a matroid (S, T) with set of bases B. Let $e \in S$. e is not in any basis if and only if e is a loop.

Proof. From the definition of a loop, e is a loop if e is in a dependent set of cardinality 1. In other words, $\{e\}$ is a dependent set. In this case, there is no independent set that contains e. Since all bases ar independent sets, no basis contains e.

Thus, if e is a loop, e is not in any basis.

Next, assume that there is an element e that is not in any basis. The independent sets of the matroid are the collection of subsets of the basis sets. Since e is not present in any basis, there are no independent sets containing e. This means that $\{e\}$ is not an independent set, so it must be a dependent set. In this case, we have a dependent set with cardinality 1, so e must be a loop.

Thus, if e is not in any basis, then e is a loop.

Putting these expressions together achieves the desired result. e is not in any basis if and only if e is a loop. ∎

3.1.15 PROBLEMS

1. Show that $B = \{\{a\}, \{b\}\}$ forms a basis for a matroid on the set $S = \{a, b\}$.
2. Show that
   $B = \{\{a, b, c, d\}, \{a, b, c, e\}, \{a, b, d, e\}, \{a, c, d, e\}, \{b, c, d, e\}\}$ forms a basis for a matroid on the set $S = \{a, b, c, d, e, f\}$.
3. Find the set of isthmuses and the set of loops for the matroid in problem 1.
4. Find the set of isthmuses and the set of loops for the matroid in problem 2.

3.2 Rank

In section 3.1 we showed that all maximal members of the independent set of a matroid have the same dimension. This unique dimension is called the rank of the matroid. More generally, this section explores the characteristics of rank and examines the rank function for a matroid.

3.2.1 DEFINITION: RANK

> Let M be a matroid (S, T), and let B be the set of all bases of M. For every pair of elements $B_1, B_2 \in B$, $|B_1| = |B_2|$. The rank of M is the dimension of the bases in M.

3.2.2 EXAMPLE

Find the rank for the matroid $M = (S, T)$ where

$$S = \{a, b, c\}$$

$$T = \{\emptyset, \{a\}, \{b\}, \{c\}, \{a, b\}, \{a, c\}\}.$$

The maximally independent sets of T are $B = \{\{a, b\}, \{a, c\}\}$. By definition, the maximally independent sets are the bases of M. We see

$$|\{a, b\}| = 2 \quad |\{a, c\}| = 2$$

so all of the bases have the same dimension as expected. The rank of M is the dimension of the bases, so $r = 2$.

3.2.3   LEMMA

The rank $r(M)$ of a matroid M satisfies

$$0 \le r(M) \le |M|.$$

Proof. The rank of M is the dimension of any basis of M. M always has at least one basis, and the dimension of the smallest possible basis is $|\emptyset| = 0$.

Alternatively, the largest possible basis of M is $|S|$. So the rank of M must be on the range $0 \le r(M) \le |M|$. ∎

3.2.4   LEMMA

Let $M = (S, T)$ be a matroid. If X is a restriction matroid of M, then

$$0 \le r(X) \le r(M).$$

Proof. Since X is a restriction matroid of M, $X \subseteq S$. The independent sets of the restriction matroid are a subset of the independent sets of M. However, the maximally independent sets in M have dimension$d \le r(M)$. Because of this, the independent sets of X may not have dimension larger than $r(M)$. This gives the desired result.∎

3.2.5   THEOREM

Let $M = (S, T)$ be a matroid and let X and Y be restriction matroids of M. Then

$$r(X \cup Y) + r(X \cap Y) \le r(X) + r(Y).$$

Proof. Designate $B_{X \cup Y}$ and $B_{X \cap Y}$ as a basis for the restriction matroids $M|X \cup Y$ and $M|X \cap Y$ respectively. Now, $|B_{X \cup Y} \cap X| \le r(X)$ and $|B_{X \cup Y} \cap Y| \le r(Y)$. So,

$$r(X) + r(Y) \ge |B_{X \cup Y} \cap X| + |B_{X \cup Y} \cap Y|$$

$$= |(B_{X \cup Y} \cap X) \cup (B_{X \cup Y} \cap Y)| + |(B_{X \cup Y} \cap X) \cap (B_{X \cup Y} \cap Y)|$$

$$= |B_{X \cup Y} \cap (X \cup Y)| + |B_{X \cup Y} \cap X \cap Y|.$$

However, $|B_{X \cup Y} \cap (X \cup Y)| = |B_{X \cup Y}| = r(X \cup Y)$, and $|B_{X \cup Y} \cap X \cap Y| = |B_{X \cap Y}| = r(X \cap Y)$, so

$$r(X) + r(Y) \ge r(X \cup Y) + r(X \cap Y). \quad \blacksquare$$

3.2.6   COROLLARY

Let $M = (S, T)$ be a matroid and let $X \subseteq Y \subseteq S$ then $r(Y) \ge r(X)$.

Proof. Let B_X be the basis of the restriction matroid $M|X$, and B_Y be the basis of the restriction matroid $M|Y$. Since $X \subset Y$, then every independent set of $M|X$ is also an independent set of $M|Y$. Thus, every basis in B_X is an independent set of $M|Y$. Moreover, the cardinality of the bases in B_X is $r(X)$.

Because a basis is a maximal independent set, the bases in B_Y must have cardinality at least as large as any independent set in $M|Y$. Furthermore, since the rank of $M|Y$ is the cardinality of the bases in $M|Y$, we know that $r(Y) \le r(X)$. $\quad \blacksquare$

3.2.7   THEOREM

Let $M = (S, T)$ be a matroid with rank function r. Let $X \subseteq S$, and let $y_1, y_2, \dots, y_n \in S$ be distinct elements such that $r(X) = r(X \cup y_1) = r(X \cup y_2) = \dots = r(X \cup y_n)$. Then $r(X \cup y_1 \cup y_2 \cup \dots \cup y_n) = r(X)$.

Proof. We prove this by induction. First, from Corollary 3.2.6, since $X \subseteq X \cup y_1 \cup y_2 \cup \dots \cup y_k$, then $r(X \cup y_1 \cup y_2 \cup \dots \cup y_k) \ge r(X)$.

For the inductive proof, we first need to prove an initial case. Let

$$Y_k = \bigcup_{i=1}^{k} y_i.$$

For $k = 1$, we have $r(X) = r(X \cup Y_1) = r(X \cup y_1)$.

Now, assume that $r(X) = r(X \cup Y_k)$ for some k. Using Theorem 3.2.5 setting X and Y from the theorem to $X = X \cup y_{k+1}$ $Y = X \cup Y_k$,

$$r\big((X \cup y_{k+1}) \cup (X \cup Y_k)\big) + r\big((X \cup y_{k+1}) \cap (X \cup Y_k)\big)$$
$$\leq r\big((X \cup y_{k+1})\big) + r(X \cup Y_k)$$

or,

$$r(X \cup Y_{k+1}) + r(X) \leq r(X) + r(X).$$

Subtracting $r(X)$ from both sides,

$$r(X \cup Y_{k+1}) \leq r(X).$$

But we showed earlier that $r(X \cup Y_k) \geq r(X)$. In order for both inequalities to be true, we must have $r(X \cup Y_{k+1}) = r(X)$.

Therefore, $r(X \cup Y_1) = r(X)$ and if $r(X \cup Y_k) = r(X)$ then we have $r(X \cup Y_{k+1}) = r(X)$, completing the inductive proof . ∎

3.2.8   EXAMPLE

Show that the cycle matroid based on Figure 15 satisfies 3.2.5, where X are Y are taken as the cycles.

The graph in the figure has two cycles:

$X = \{a, b, c, d\}$ and $Y = \{a', b', c', d'\}$.

The maximally independent sets of the matroid are:

$\{a,b,c,e,a',b',c'\}$ $\{a,b,c,e,a',b',d'\}$ $\{a,b,c,e,a',c',d'\}$ $\{a,b,c,e,b',c',d'\}$

$\{a,b,d,e,a',b',c'\}$ $\{a,b,d,e,a',b',d'\}$ $\{a,b,d,e,a',c',d'\}$ $\{a,b,d,e,b',c',d'\}$

$\{a,c,d,e,a',b',c'\}$ $\{a,c,d,e,a',b',d'\}$ $\{a,c,d,e,a',c',d'\}$ $\{a,c,d,e,b',c',d'\}$

$\{b,c,d,e,a',b',c'\}$ $\{b,c,d,e,a',b',d'\}$ $\{b,c,d,e,a',c',d'\}$ $\{b,c,d,e,b',c',d'\}$

The union and intersection of X and Y are:

$$X \cup Y = \{a,b,c,d,a',b',c',d'\} \quad X \cap Y = \emptyset$$

The restriction matroid $M|(X \cup Y)$ has bases:

$\{a,b,c,a',b',c'\}$ $\{a,b,c,a',b',d'\}$ $\{a,b,c,a',c',d'\}$ $\{a,b,c,b',c',d'\}$

$\{a,b,d,a',b',c'\}$ $\{a,b,d,a',b',d'\}$ $\{a,b,d,a',c',d'\}$ $\{a,b,d,b',c',d'\}$

$\{a,c,d,a',b',c'\}$ $\{a,c,d,a',b',d'\}$ $\{a,c,d,a',c',d'\}$ $\{a,c,d,b',c',d'\}$

$\{b,c,d,a',b',c'\}$ $\{b,c,d,a',b',d'\}$ $\{b,c,d,a',c',d'\}$ $\{b,c,d,b',c',d'\}$

So the rank of $M|(X \cup Y)$ is 6.

Alternatively, the bases of the restriction matroid $M|(X \cap Y)$ is simply $B_{X \cap Y} = \{\emptyset\}$, so the rank of $M|(X \cap Y)$ is 0.

Furthermore, the restriction matroid $M|X$ has bases:

$$\{a,b,c\} \quad \{a,b,d\}$$
$$\{a,c,d\} \quad \{b,c,d\}$$

giving the rank of $M|X$ as 3.

Similarly, the restriction matroid $M|Y$ has bases:

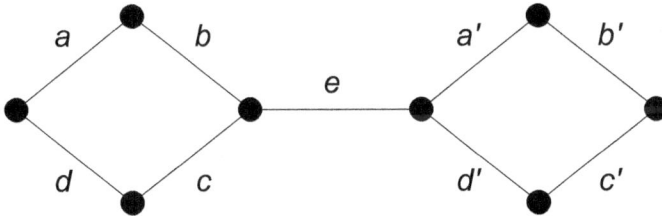

Figure 15: Bridge graph.

$$\{a',b',c'\} \quad \{a',b',d'\}$$
$$\{a',c',d'\} \quad \{b',c',d'\}$$

giving the rank of $M|Y$ also as 3.

Testing theorem 3.2.5,

$$r(X \cup Y) + r(X \cap Y) \le r(X) + r(Y)$$

$$6 + 0 \le 3 + 3$$

in line the results of the theorem.

3.2.9 LEMMA

The rank of a matroid M arising from a connected tree graph G is $r(M) = V - 1$, where V is the number of vertices of the graph.

Proof. The bases of the matroid correspond to the spanning trees of the graph. Euler's formula for graphs states,

$$V - E + F = 2,$$

where V is the number of vertices, E is the number of edges, and F is the number of faces. A tree has F=1 (remember that the outer, infinitely large region of the graph counts as a face), so $E = V - 1$ for a tree.

The bases of M are all spanning trees with the same number of vertices as in the original graph G. The number of edges present in each basis is $E = V - 1$. Thus, $|B| = V - 1$ for all the bases of M, and since the rank of M is the dimension of the bases, we have $r(M) = V - 1$. ∎

3.2.10 LEMMA

The rank of a matroid M arising from a matrix A is the same as the rank of A.

Proof. The basis of M is the formed from the largest sets of linear independent column vectors in A. However, the rank of A is the maximum number of linearly independent columns in A. Thus, the rank of M is equal to the rank of A. ∎

3.2.11 LEMMA

Let M be a matroid with rank $r(M)$ and let t be an independent set of M where $|t| = r(M)$. Then t is a basis of M.

Proof. From the definition of rank, we know that all bases b for M have $|b| = r(M)$. If t is an independent set that is not a basis, then t is not a maximal independent set (otherwise t would be a basis by definition). If t is not a maximal independent, then t is a subset of another independent set. If this is the case, then there is an independent set of M such that $t \subseteq s$ where $|s| > |t| = r(M)$. However, in this situation, the rank of M must be at least $|s|$, which contradicts the assertion that the rank of M is $r(M)$. Thus, t must be a maximal independent set, and therefore t is a basis of M. ∎

3.2.12 THEOREM

Let M be a matroid. An independent set t is a basis of M if and only if $|t| = r(M)$.

Proof. From the definition of rank, if b is a basis of M, then $|b| = r(M)$. Moreover, lemma 3.2.9 showed that if $|b| = r(M)$, then b is a basis of M.∎

3.2.13 DEFINITION: SUBSET RANK

Let M be a matroid (S, T), and let $A \subseteq S$ be a subset of S. The rank of A in M is the rank of the restriction matroid $M|A$.

3.2.14 EXAMPLE

Let $M = (S, B)$ be the matroid over the set S with set of bases B such that

$$S = \{a, b, c, d\}$$

$$B = \{\{a, b, c, d\}\}$$

Let $A = \{a, b\}$. Find the rank of A.

Since $\{a, b, c, d\}$, every subset of S is an independent set. The set A is an independent set, and this is also the maximal independent set in A. Thus the rank of A is $r(A) = 2$.

3.2.15 EXAMPLE

Let $M = (S, B)$ be the matroid over the set S with set of bases B such that

$$S = \{a, b, c, d\}$$

$$B = \{\{a\}, \{b\}\}$$

Let $A = \{a, b\}$. Find the rank of A.

The largest independent sets that are subsets of a are $\{a\}$ and $\{b\}$. Given this, the rank of A is $r(A) = 1$.

3.2.16 PROPOSITION

Let M be a matroid with ground set S and rank function r. Let $X \subseteq S$ be a subset of the ground set. Then

RP-1   X is an independent set if and only if $|X| = r(X)$

RP-2   X is a basis if and only if $|X| = r(X) = r(M)$

Proof. First, if I is an independent set of M then the restriction matroid $M|I$ will have I as one of its independent sets. Moreover, I is the largest independent set in $M|I$ because all independent sets in $M|I$ are subsets of I. Since the rank of I in M is the rank of the restriction $M|I$,

$$r_M(I) = r_{M|I}(M|I) = |I|.$$

Thus, if I is an independent set of M, then $|I| = r_M(I)$.

Now, given a S set such that $r_M(S) = |S|$, we know that $r_{M|S}(M|S) = |S|$. But for this to be true, S must be an independent set in $M|S$. Also, every independent set in $M|S$ is also an independent set in M, so S is also an independent set of M.

Therefore, if $|S| = r_M(S)$, then S is an independent set of M.

Combining these two statements:

X is an independent set of M if and only if $|X| = r(X)$.

As far as the second statement, we know that if X is a basis then $|X| = r(M)$ because the rank of M is defined as the cardinality of the basis elements.

So, if X is a basis then $|X| = r(M)$.

Furthermore, if S is a set such that $|S| = r_M(S) = r_M(M)$, then S must be an independent set in $M|S$, which means it is an independent set of M as well. But S has the same cardinality as the basis elements of M. If S is an independent set of M with cardinality $r_M(M)$, then S must be a basis. Otherwise, S must be a subset of some basis element B, and all subsets have cardinality smaller than their parent set, unless $S = B$.

This means that if $|S| = r_M(S) = r_M(M)$, then S is a basis of M.

From the last two statements we know that:

X is a basis of and only if $|X| = r(X) = r(M)$. ■

3.2.17 DEFINITION: RANK MATROID

> Let S be a set, $X \subseteq Y \subseteq S$, and $r(X)$ a function satisfting
>
> **RM-1** $0 \le r(X) \le |X|$
>
> **RM-2** $r(X) \le r(Y)$
>
> **RM-3** $r(X \cup Y) + r(X \cap Y) \le r(X) + r(Y)$
>
> Let $\Im$ be the collection of subsets of S such that $r(X) = |X|$. Then the ordered pair $(S, \Im)$ forms a matroid.

Proof. We show that the above definition satisfies **SM-1**, **SM-2**, and **SM-3**.

SM-1 $\emptyset \in \mathfrak{I}$

First, $\emptyset \subseteq S$. Moreover, from **RM-1**, $0 \leq r(\emptyset) \leq |\emptyset| = 0$. Thus, $r(\emptyset) = 0$, which means that $\emptyset \in \mathfrak{I}$.

SM-2   If $t \in \mathfrak{I}$ and $\bar{t} \subseteq t$, then $\bar{t} \in \mathfrak{I}$

From **SM-3**, we see

$$r\big(\bar{t} \cup (t - \bar{t})\big) + r\big(\bar{t} \cap (t - \bar{t})\big) \leq r(\bar{t}) + r(t - \bar{t}).$$

In the above expression, $\bar{t} \cup (t - \bar{t}) = t$, and $\bar{t} \cap (t - \bar{t}) = \emptyset$. Substituting this into the above inequality,

$$r(t) + r(\emptyset) \leq r(\bar{t}) + r(t - \bar{t}).$$

Since $t \in \mathfrak{I}, r(t) = |t|$ and $r(\emptyset) = 0$,

$$|t| + 0 \leq r(\bar{t}) + r(t - \bar{t}).$$

From **RM-2**, $r(\bar{t}) \leq |\bar{t}|$ and $r(t - \bar{t}) \leq |t - \bar{t}|$. Substituting this into the right side of the inequality,

$$|t| \leq r(\bar{t}) + r(t - \bar{t}) \leq |\bar{t}| + |t - \bar{t}|$$

But since $\bar{t} \cup (t - \bar{t}) = t$, and $\bar{t} \cap (t - \bar{t}) = \emptyset$, $|\bar{t}| + |t - \bar{t}| = |t|$. Thus,

$$|t| \leq r(\bar{t}) + r(t - \bar{t}) \leq |t|.$$

Because $r(\bar{t}) \leq |\bar{t}|$ and $r(t - \bar{t}) \leq |t - \bar{t}|$, the only way the above inequality can hold is if $r(\bar{t}) = |\bar{t}|$ and $r(t - \bar{t}) = |t - \bar{t}|$.

If $r(\bar{t}) = |\bar{t}|$ and $r(t - \bar{t}) = |t - \bar{t}|$, then both $\bar{t}$ and $t - \bar{t}$ are members of $\mathfrak{I}$. Thus, if $t \in \mathfrak{I}$ and $\bar{t} \subseteq t$, then $\bar{t} \in \mathfrak{I}$.

SM-3   Let $t_1, t_2 \in T$ where $|t_1| < |t_2|$, then $\exists\, d \in t_2 - t_1$ such that $t_1 \cup d \in T$

Assume that **SM-3** does not hold. Let $t_1, t_2 \in T$ with $|t_1| < |t_2|$, and let $d \in t_2 - t_1$. If **SM-3** does not hold then from **RM-1**, $r(t_1 \cup d) < |t_1 \cup d|$.

Because $t_1 \subseteq t_1 \cup d$, from **RM-2** we know $r(t_1 \cup d) \geq r(t_1) = |t_1|$. Combining these inequalities,

$$|t_1| \leq r(t_1 \cup d) < |t_1 - d|.$$

However, it is impossible for $|t_1| < |t_1 - d|$, invalidating the assumption that **SM-3** does not hold.

Thus, we see Definition 3.2.17 implies **SM-1**, **SM-2**, and **SM-3**. Thus, the ordered pair $(S, \mathfrak{I})$ forms a matroid. ∎

3.2.18 DEFINITION: NULLITY

> Let M be a matroid (S, T), let B be the set of all bases of M, let r be the rank function on M, and let $X \subseteq S$ be a subset of the ground set. The nullity of X is
>
> $$n_M(X) = |X| - r(X)$$

3.2.19 EXAMPLE

Find the nullity for the sets in examples 3.2.12and 3.2.13.

In both cases the set is $A = \{a, b\}$ so $|A| = 2$. For the first case we have

$$n_{M_1}(A) = |A| - r(A) = 2 - 2 = 0$$

For the second case,

$$n_{M_2}(A) = |A| - r(A) = 2 - 1 = 1$$

3.2.20 PROBLEMS

1. Find the rank of the matroid $M = (S, B)$
$$S = \{a, b, c, d\}$$
$$B = \{\{a, b, d\}, \{a, c, d\}. \{b, c, d\}\}$$

2. With M from problem 1, find the rank of the matroids:
   a. $M|A$ where $A = \{a, b, c\}$
   b. $M|A$ where $A = \{a, b, d\}$
   c. $M|A$ where $A = \{b, d\}$
   d. $M|A$ where $A = \{c\}$

3. Find the nullity of the subsets A in problem 2.

4. Find the rank of the matroid $M = (S, B)$
$$S = \{a, b, c, d, e\}$$
$$B = \{\{a, c\}, \{a, d\}. \{b, c\}, \{b, d\}\}$$

5. With M from problem 1, find the rank of the matroids:
   a. $M|A$ where $A = \{a, b, e\}$
   b. $M|A$ where $A = \{a, b, c\}$
   c. $M|A$ where $A = \{b, d\}$
   d. $M|A$ where $A = \{c, d\}$
   e. $M|A$ where $A = \{a\}$
   f. $M|A$ where $A = \{e\}$

6. Find the nullity of the subsets A in problem 2.

3.3   Closure

In the theory of vector spaces, the span of a set of vectors gives a subspace generated by the vectors. The matroid equivalent is the closure of the matroid.

3.3.1   DEFINITION: CLOSURE

Let M be a matroid (S, T), and let $M|X$ be a restriction matroid. The closure of X is the set

$$cl(X) = \{x \in S \mid r(X \cup x) = r(X)\}.$$

3.3.2   EXAMPLE

Let M be a matroid (S, T) with $S = \{a, b, c, d, e\}$ and basis set $B = \{\{a, b, d\}, \{b, d, e\}\}$. Find the closure of M with respect to $\emptyset$, $\{a\}$, $\{c\}$, $\{a, b\}$, $\{b, d\}$, $\{b, e\}$, and $\{b, d, c\}$.

From the basis set we know $r(M) = 3$. To find the closure for the sets in question, it will be useful to list all independent sets of M:

3-member independent sets:

$$\{a, b, d\} \quad \{b, d, e\}$$

2-member independent sets:

$$\{a, b\} \quad \{a, d\} \quad \{b, d\}$$

$$\{b, e\} \quad \{d, e\}$$

1-member independent sets:

$$\{a\} \quad \{b\} \quad \{d\} \quad \{e\}$$

0-member independent sets:

$$\emptyset$$

$X = \emptyset$

To find the closure, we first find the restriction matroid $M|X$, then we find the rank of $X \cup e_i$ for every $e_i \in S$. If any of these have the same rank as $M|X$, then e_i belongs in the closure.

The matroid $M|X$ has bases set $B = \{\emptyset\}$, so $r(M|X) = 0$.

$$X \cup \{a\} = \{a\} \quad r(M|\{a\}) = 1$$

$$X \cup \{b\} = \{b\} \quad r(M|\{b\}) = 1$$

$$X \cup \{c\} = \{c\} \quad r(M|\{c\}) = 0$$

$$X \cup \{d\} = \{d\} \quad r(M|\{d\}) = 1$$

$$X \cup \{e\} = \{e\} \quad r(M|\{e\}) = 1$$

Only the element c has the desired property, so $cl(\emptyset) = \{c\}$. All other elements increase the rank of the matroid when joined with the original set X.

$X = \{a\}$

The matroid $M|X$ has bases set $B = \{\{a\}\}$, so $r(M|X) = 1$.

$$X \cup \{a\} = \{a\} \quad r(M|\{a\}) = 1$$

$$X \cup \{b\} = \{a, b\} \quad r(M|\{a, b\}) = 2$$

$$X \cup \{c\} = \{a, c\} \quad r(M|\{a, c\}) = 1$$

$$X \cup \{d\} = \{a, d\} \quad r(M|\{a, d\}) = 2$$

$$X \cup \{e\} = \{a, e\} \quad r(M|\{a, e\}) = 2$$

Elements a and c have the desired properties, so $cl(\{a\}) = \{a, c\}$.

$X = \{c\}$

The matroid $M|X$ has bases $B = \{\emptyset\}$ (there are no independent sets that contain element c), so $r(M|X) = 0$.

$$X \cup \{a\} = \{a\} \quad r(M|\{a\}) = 1$$

$$X \cup \{b\} = \{b\} \quad r(M|\{b\}) = 1$$

$$X \cup \{c\} = \{c\} \quad r(M|\{c\}) = 0$$

$$X \cup \{d\} = \{d\} \quad r(M|\{d\}) = 1$$

$$X \cup \{e\} = \{e\} \quad r(M|\{e\}) = 1$$

Elements a and c have the desired properties, so $cl(\{c\}) = \{c\}$.

$X = \{a, b\}$

The matroid $M|X$ has bases set $B = \{\{a, b\}\}$, so $r(M|X) = 2$.

$$X \cup \{a\} = \{a, b\} \quad r(M|\{a, b\}) = 2$$

$$X \cup \{b\} = \{a, b\} \quad r(M|\{a, b\}) = 2$$

$$X \cup \{c\} = \{a, b, c\} \quad r(M|\{a, b, c\}) = 2$$

$$X \cup \{d\} = \{a, b, d\} \quad r(M|\{a, b, d\}) = 3$$

$$X \cup \{e\} = \{a, b, e\} \quad r(M|\{a, b, e\}) = 2$$

Therefore, $cl(\{a, b\}) = \{a, b, c, e\}$.

$X = \{b, d\}$

The matroid $M|X$ has bases $B = \{\{b, d\}\}$, so $r(M|X) = 2$.

$$X \cup \{a\} = \{a, b, d\} \quad r(M|\{a, b, d\}) = 3$$

$$X \cup \{b\} = \{b, d\} \quad r(M|\{b, d\}) = 2$$

$$X \cup \{c\} = \{b, c, d\} \quad r(M|\{b, c, d\}) = 2$$

$$X \cup \{d\} = \{b, d\} \quad r(M|\{b, d\}) = 2$$

$$X \cup \{e\} = \{b, d, e\} \quad r(M|\{b, d, e\}) = 2$$

Therefore, $cl(\{b, d\}) = \{b, c, d, e\}$.

$X = \{b, e\}$

The matroid $M|X$ has bases $B = \{\{b\}, \{e\}\}$, so $r(M|X) = 1$.

$$X \cup \{a\} = \{a, b, e\} \quad r(M|\{a, b, d\}) = 2$$

$$X \cup \{b\} = \{b, e\} \quad r(M|\{b, d\}) = 1$$

$$X \cup \{c\} = \{b, c, e\} \quad r(M|\{b, c, e\}) = 1$$

$$X \cup \{d\} = \{b, d, e\} \quad r(M|\{b, d, e\}) = 2$$

$$X \cup \{e\} = \{b, e\} \quad r(M|\{b, e\}) = 1$$

Therefore, $cl(\{b, e\}) = \{b, c, e\}$.

$X = \{b, c, d\}$

The matroid $M|X$ has bases set $B = \{\{b, d\}\}$, so $r(M|X) = 2$.

$$X \cup \{a\} = \{a, b, c, d\} \quad r(M|\{a, b, c, d\}) = 3$$

$$X \cup \{b\} = \{b, c, d\} \quad r(M|\{b, c, d\}) = 2$$

$$X \cup \{c\} = \{b, c, d\} \quad r(M|\{b, c, d\}) = 2$$

$$X \cup \{d\} = \{b, c, d\} \quad r(M|\{b, c, d\}) = 2$$

$$X \cup \{e\} = \{b, c, d, e\} \quad r(M|\{b, c, d, e\}) = 3$$

Therefore, $cl(\{b, c, d\}) = \{b, c, d\}$.

3.3.3   LEMMA

Let M be a matroid (S, T) and let $X \subseteq S$. Then $X \subseteq cl(X)$.

Proof. Since $X \cup x = X$ for every $x \in X$, by definition, the closure of X must contain each of these x's. Thus, $X \subseteq cl(X)$. ∎

3.3.4   LEMMA

Let M be a matroid (S, T) and let $X \subseteq S$. Then $r(X) = r\big(cl(X)\big)$.

Proof. From Lemma 3.3.3, we know $X \subseteq cl(X)$. If $X = cl(X)$, the result follows immediately.

Assume $X \neq cl(X)$. Let $c_1 \in cl(X) - X$. From the definition of closure, $r(X \cup c_1) = r(X)$. If $|cl(X) - X| = 1$, then c_1 is the only element of $cl(X) - X$, so $X \cup c_1 = cl(X)$, and we have $r(X) = r(X \cup c_1) = r\big(cl(X)\big)$.

Otherwise, $|cl(X) - X| > 1$, and there is another element $c_2 \in cl(X) - X$ where $c_2 \neq c_1$. Again, from the definition of closure we know $r(X) = r(X \cup c_2)$.

From Theorem 3.2.5, we have

$$r\big((X \cup c_1) \cup (X \cup c_2)\big) + r\big((X \cup c_1) \cap (X \cup c_2)\big)$$
$$\leq r(X \cup c_1) + r(X \cup c_2).$$

Because $c_2 \neq c_1$ and $c_1, c_2 \notin X$, we know $(X \cup c_1) \cup (X \cup c_2) = X \cup c_1 \cup c_2$, and $(X \cup c_1) \cap (X \cup c_2) = X$. Substituting this into the inequality and noting that $r(X \cup c_1) = r(X \cup c_2) = r(X)$,

$$r(X \cup c_1 \cup c_2) + r(X) \leq 2r(X)$$

or

$$r(X \cup c_1 \cup c_2) \leq r(X).$$

But from Corollary 3.2.6, $r(X \cup c_1 \cup c_2) \geq r(X)$. This only leaves the option $r(X \cup c_1 \cup c_2) = r(X)$.

We can repeat this process for every element in $cl(X) - X$. This leads to

$$r(X) = r(X \cup c_1 \cup c_2 \cup ... \cup c_n) = r\big(X \cup (cl(X) - X)\big) = r(cl(X)). ∎$$

3.3.5   THEOREM

Let M be a matroid (S, T) and let $X \subseteq S$. Then $cl\big(cl(X)\big) = cl(X)$.

Proof. Let $Y = cl(X)$. Assume $Y \neq cl(Y)$. We know from Lemma 3.3.3 that $Y \subseteq cl(Y)$, and the assumption that $Y \neq cl(Y)$ means $Y \subset cl(Y)$.

In this case, the set $Z = cl(Y) - Y$ is not empty. Let $z \in Z$ be an element of Z. Because $Y \subset cl(Y)$, z is an element of $cl(Y)$ but not an element of Y. Symbolically, $z \in cl(Y)$ but $z \notin Y$.

From the definition of closure, $r(Y \cup z) = r(Y)$. But since $Y = cl(X)$, from Lemma 3.3.4 we know $r(Y) = r\big(cl(X)\big) = r(X)$, so we have $r(Y \cup z) = r(X)$.

From Corollary 3.2.6 we know $r(X \cup z) \geq r(X)$. Similarly, because $X \subseteq Y$, we have $X \cup z \subseteq Y \cup z$, so $r(X \cup z) \leq r(Y \cup z)$. Putting these together,

$$r(Y \cup z) \geq r(X \cup z) \geq r(X).$$

But we have already shown that $r(Y \cup z) = r(X)$, so it must be the case that $r(Y \cup z) = r(X \cup z) = r(X)$.

However, if $r(X \cup z) = r(X)$, then $z \in cl(X) = Y$. But this contradicts our assumption that $z \notin Y$. Thus, no such z exists, and our initial assumption that $z \in cl(Y) - Y$ is invalid.

Because there are no elements in $cl(Y) - Y$, and $Y \subseteq cl(Y)$, we have $Y = cl(Y)$ which is equivalent to $cl(X) = cl\big(cl(X)\big)$. ∎

3.3.6   DEFINITION: FLAT

Let M be a matroid (S, T) and let $X \subseteq S$ be a subset of the ground set of M. X is a flat of M if

$$X = cl(X).$$

3.3.7   LEMMA

Let M be a matroid (S, T) and let $X \subseteq S$ be a subset of the ground set of M. The closure $cl(X)$ is a flat.

Proof. Set $Y = cl(X)$. From Theorem 3.3.5, $cl(Y) = cl(cl(X)) = cl(X) = Y$. Since $Y = cl(Y)$, Y is a flat, so $cl(X)$ is a flat.∎

3.3.8   PROPOSITION

Let M be a matroid with ground set S. The closure operator for M satisfies

CL-1   If $X \subseteq S$ then $X \subseteq cl(X)$

CL-2   If $X \subseteq Y \subseteq S$ then $cl(X) \subseteq cl(Y)$

CL-3   If $X \subseteq S$ then $cl(cl(X)) = cl(X)$

CL-4   If $X \subseteq S$, $y \in S$, and $z \in cl(X \cup y) - cl(X)$, then $y \in cl(X \cup z)$

3.3.9   PROPOSITION

Let S be a set, cl be an operator on S satisfying **CL-1-4**, and let $\mathfrak{I}$ be a set defined as

$$\mathfrak{I} = \{A \subseteq S \mid a \notin cl(A - a) \, \forall \, a \in A\}.$$

Furthermore, let $X \in \mathfrak{I}$ and $y \in S$. If $X \cup y \notin \mathfrak{I}$ then $y \in cl(X)$.

Proof. If $X \cup y \notin \mathfrak{I}$, then there is some $z \in X$ such that $z \in cl(X \cup y - z)$. We divide the proof into two cases: $y \neq z$ and $y = z$.

If $y = z$ then the proposition is satisfied because then $y \in cl(X - y)$. But $y \notin X$ because $X \in \mathfrak{I}$ which means that every element $x \in X$ satisfies $x \notin cl(X - x)$. However, if $y \notin X$, then $X - y = X$, so $y \in cl(X)$.

If $y \neq z$ then $X \cup y - z = (X - z) \cup y$. Moreover, since $X \in \mathfrak{I}$ and $z \in X$, then $z \notin cl(X - z)$. Putting these together, $cl(X \cup y - z) = cl((X - z) \cup y)$, and since $z \in cl(X \cup y - z)$, then $z \in$

$cl\big((X-z)\cup y\big)$. However, since $z\notin cl(X-z)$ we also know $z\in cl\big((X-z)\cup y\big)-cl(X-z)$.

But if $z\in cl\big((X-z)\cup y\big)-cl(X-z)$, then we can invoke **CL-4** which gives $y\in cl\big((X-z)\cup z\big)$. But since $(X-z)\cup z=X$, we have $y\in cl(X)$.

Thus, in either case, If $X\cup y\notin\Im$ then $y\in cl(X)$.■

3.3.10 DEFINITION: CLOSURE MATROID

Let S be a set and let $\Im$ be the set

$$\Im=\{A\subseteq S\mid a\notin cl(A-a)\ \forall\ a\in A\}$$

where the function $cl(X)$ satisfies

CL-1   If $X\subseteq S$ then $X\subseteq cl(X)$

CL-2   If $X\subseteq Y\subseteq S$ then $cl(X)\subseteq cl(Y)$

CL-3   If $X\subseteq S$ then $cl\big(cl(X)\big)=cl(X)$

CL-4   If $X\subseteq S,\ y\in S$, and $z\in cl(X\cup y)-cl(X)$, then $y\in cl(X\cup z)$

Then the ordered pair $(S,\Im)$ forms a matroid.

Proof. We show that the ordered pair $(S,\Im)$ as defined satisfy the axioms **SM-1**, **SM-2**, and **SM-3** of a set matroid.

SM-1 $\emptyset\in\Im$

Let $A=\emptyset$ in the definition of $\Im$. $A\subseteq S$ because $\emptyset$ is a subset of S. But since there are no elements to A, it is trivially true that $a\notin cl(A-a)\ \forall\ a\in A$. Thus $\emptyset\in\Im$.

SM-2   If $T\in\Im$ and $\bar T\subseteq T$, then $\bar T\in\Im$

If $T \in \Im$, from the definition of $\Im$ we know that $a \notin cl(T - a) \; \forall \; a \in T$. Let $t \in \bar{T}$. Since $\bar{T} \subseteq T$, we know $t \in T$ which means that $t \notin cl(T - t)$.

Because $\bar{T} \subseteq T, \bar{T} - t \subseteq T - t$. But from **CL-2**, we have $cl(\bar{T} - t) \subseteq cl(T - t)$. So if $t \notin cl(T - t)$, then $t \notin cl(\bar{T} - t)$.

This is true for every $t \in \bar{T}$. Thus, $t \notin cl(\bar{T} - t) \; \forall \; t \in \bar{T}$, which means $\bar{T} \in \Im$ from the definition of $\Im$.

SM-3   Let $T_1, T_2 \in \Im$ where $|T_1| < |T_2|$, then $\exists \; d \in T_2 - T_1$ such that $T_1 \cup d \in \Im$

We prove this by assuming that **SM-3** fails, and finding a contradiction. If **SM-3** fails, then identify all pairs (T_1, T_2) where **SM-3** is not satisfied. From this group, choose a pair where $|T_1 \cap T_2|$ is a maximum. Pick a $y \in T_2 - T_1$.

We will need the result of Proposition 3.3.9 in a slightly different form. We can formulate 3.3.9 as the logical statement

$$A \wedge B \to C$$

where $A = X \in \Im$, $B = X \cup y \notin \Im$, and $C = y \in cl(X)$. Now, if a statement is true, then its contrapositive is also true. The contrapositive to our statement is

$$\bar{C} \to \overline{A \wedge B}$$

where the overbars indicate the negation of the statement. But $\overline{A \wedge B} = \bar{A} \vee \bar{B}$. Thus

$$\bar{C} \to \bar{A} \vee \bar{B}.$$

Which is equivalent to

$$\bar{C} \wedge A \to \bar{B}$$

Now, $\bar{C} = y \notin cl(X)$, and $\bar{B} = X \cup y \in \Im$. Putting this into the new statement, our reformulated version of Proposition 3.3.9 reads

If $X \in \Im$ and $y \notin cl(X)$ then $X \cup y \in \Im$.

We divide the proof into two cases: $T_1 \subseteq cl(T_2 - y)$ and $T_1 \not\subseteq cl(T_2 - y)$. We show both cases lead to a contradiction.

$T_1 \subseteq cl(T_2 - y)$

In this case, we know from **CL-3** that $cl(T_1) \subseteq cl(T_2 - y)$. But since $T_2 \in \mathfrak{I}$, we know $y \notin cl(T_2 - y)$. Thus, $y \notin cl(T_1)$. But from our reformulated Proposition 3.3.9, if $T_1 \in \mathfrak{I}$ and $y \notin cl(T_1)$, then $T_1 \cup y \in \mathfrak{I}$.

But this would mean that **SM-3** does not fail which is contradictory to our initial assumption.

$T_1 \not\subseteq cl(T_2 - y)$

In this case, there must be some element $z \in T_1$ where $z \notin cl(T_2 - y)$. We begin by showing that $z \notin T_2$. First, $z \notin T_2 - y$ because by **CL-1**, every element of $T_2 - y$, must be present in $cl(T_2 - y)$, and since $z \notin cl(T_2 - y)$, then $z \notin T_2 - y$.

However, $z \neq y$ because $z \in T_1$ while $y \in T_2 - T_1$. Since , $z \notin T_2 - y$ and $z \neq y$, we have $z \notin T_2$.

Furthermore, since $T_2 \in \mathfrak{I}$, then $T_2 - y \in \mathfrak{I}$ (this is **SM-2**). But if $T_2 - y \in \mathfrak{I}$ and $z \notin cl(T_2 - y)$, then our reformulated Proposition 3.3.9 tells us that $(T_2 - y) \cup z \in \mathfrak{I}$.

However, we know that **SM-3** is valid for the pair (T_1, T_3) with $T_3 = (T_2 - y) \cup z$. This must be true because $|T_1 \cap T_3| > |T_1 \cap T_2|$, and we specifically chose the pair (T_1, T_2) because $|T_1 \cap T_2|$ is maximal among the pairs of sets where **SM-3** fails. We know that $|T_1 \cap T_3| > |T_1 \cap T_2|$ because we construct T_3 by taking T_2, removing an element y that is not part of T_1, then adding in z which is in T_1.

Now, since **SM-3** is valid for (T_1, T_3), there must be some element $d \in T_3$ such that $d \notin T_1$ and $T_1 \cup d \in \mathfrak{I}$. But since $T_3 = (T_2 - y) \cup z$, every element in T_3, except z, is also in T_2. We know that $d \neq z$ because $d \notin T_1$ while $z \in T_1$.

Thus, d must be some element in T_3 that is also in T_2. But if this is true, and if $T_1 \cup d \in \mathfrak{I}$, then we have discovered an element $d \in T_2 - T_1$ such that $T_1 \cup d \in \mathfrak{I}$. This means that **SM-3** is satisfied, which is contradictory to our assumption.

Therefore, we have shown in both cases that the assumption that SM-3 is not satisfied is false. We conclude that **SM-3** is valid. Furthermore, we note that our proof requires the definition of $\mathfrak{I}$ as well as **CL-1**, **CL-2**, **CL-3**, and **CL-4**.∎

3.3.11 EXAMPLE

Let $S = \{a, b\}$ with a function $\mathfrak{f}$ defined by the mappings

$$\mathfrak{f}(\emptyset) = \{b\} \quad \mathfrak{f}(\{a\}) = \{a, b\} \quad \mathfrak{f}(\{b\}) = \{b\} \quad \mathfrak{f}(\{a, b\}) = \{a, b\}$$

Let $\mathfrak{H} = \{\emptyset, \{a\}\}$. Show that the ordered pair $(S, \mathfrak{H})$ is a matroid.

First, we show that $\mathfrak{f}$ satisfies **CL-1-4**.

CL-1   If $X \subseteq S$ then $X \subseteq \mathfrak{f}(X)$

Examining each relation we see

$$\emptyset \subseteq \{b\} \quad \{a\} \subseteq \{a, b\} \quad \{b\} \subseteq \{b\} \quad \{a, b\} \subseteq \{a, b\}$$

so **CL-1** is satisfied in each case.

CL-2   If $X \subseteq Y \subseteq S$ then $cl(X) \subseteq cl(Y)$

Again, we inspect each case. There are four possibilities for Y, and up to four possibilities for X.

$Y = \{a, b\}$ $\mathfrak{f}(Y) = \{a, b\}$

The possible values for X are

$$\emptyset \quad \{a\} \quad \{b\} \quad \{a, b\}$$

Examining each case,

$$\mathfrak{f}(\emptyset) = \{b\} \subseteq \{a, b\} \quad \mathfrak{f}(\{a\}) = \{a, b\} \subseteq \{a, b\}$$

$$\mathfrak{f}(\{b\}) = \{b\} \subseteq \{a, b\} \quad \mathfrak{f}(\{a, b\}) \subseteq \{a, b\}$$

$Y = \{b\}$ $f(Y) = \{b\}$

The possible values for X are

$$\emptyset \quad \{b\}$$

Examining each case,

$$f(\emptyset) = \{b\} \subseteq \{b\} \quad f(\{b\}) = \{b\} \subseteq \{b\}$$

$Y = \{a\}$ $f(Y) = \{a, b\}$

The possible values for X are

$$\emptyset \quad \{a\}$$

Examining each case,

$$f(\emptyset) = \{a\} \subseteq \{a, b\} \quad f(\{a\}) = \{a, b\} \subseteq \{a, b\}$$

$Y = \emptyset$ $f(Y) = \{b\}$

The possible values for X are

$$\emptyset$$

Examining this case,

$$f(\emptyset) = \{b\} \subseteq \{b\}$$

Thus, in each case **CL-2** is satisfied.

CL-3   If $X \subseteq S$ then $f(f(X)) = f(X)$

Again, examine each of the four values for X.

$$f(f(\emptyset)) = f(\{b\}) = \{b\} = f(\emptyset) \quad f(f(\{a\})) = f(\{a, b\}) = \{a, b\} = f(\{a\})$$

$$f(f(\{b\})) = f(\{b\}) = \{b\} = f(\{b\}) \quad f(f(\{a, b\})) = f(\{a, b\}) = \{a, b\} = f(\{a, b\})$$

CL-4   If $X \subseteq S$, $y \in S$, and $z \in f(X \cup y) - f(X)$, then $y \in f(X \cup z)$

There are four possibilities for X, two possibilities for y, and up to two possibilities for z. We can eliminate the values of y where $y \in X$. In this case, $X \cup y = X$. There are no values of z to choose, so **CL-4** is trivially satisfied. Thus, we only need be concerned with $y \notin X$.

$X = \emptyset \quad y = a$

In this case, $z \in f(\emptyset \cup \{a\}) - f(\emptyset) = f(\{a\}) - f(\emptyset) = \{a, b\} - \{b\} = \{a\}$. We see $f(\emptyset \cup \{a\}) = f(\{a\}) = \{a, b\}$, and $y = a \in \{a, b\}$.

$X = \emptyset \quad y = b$

In this case, $z \in f(\emptyset \cup \{b\}) - f(\emptyset) = f(\{b\}) - f(\emptyset) = \{b\} - \{b\} = \emptyset$. There are no elements for z to choose, so **CL-4** is trivially satisfied.

$X = \{a\} \quad y = b$

In this case, $z \in f(\{a\} \cup \{b\}) - f(\{a\}) = f(\{a, b\}) - f(\{a\}) = \{a, b\} - \{a, b\} = \emptyset$. Again, there are no elements for z to choose, so **CL-4** is trivially satisfied.

$X = \{b\} \quad y = a$

In this case, $z \in f(\{b\} \cup \{a\}) - f(\{b\}) = f(\{a, b\}) - f(\{b\}) = \{a, b\} - \{b\} = \{a\}$. We have $f(\{b\} \cup \{a\}) = f(\{a, b\}) = \{a, b\}$, and $y = a \in \{a, b\}$.

In each case **CL-4** is satisfied.

Now, since f satisfies **CL-1-4**, we can interpret f as a closure operator. We still need to show that the sets $\mathfrak{H}$ satisfy the definition for $\mathfrak{I}$.

$\mathfrak{I} = \{A \subseteq S \mid x \notin cl(A - x) \; \forall \, x \in A\}$

There are four possible values for A. The definition should leed to the two sets in $\mathfrak{H}$. We need to test each of these sets to see that the definition for $\mathfrak{I}$ is satisfied.

$A = \emptyset$

Since there are no elements $x \in A$, the definition is trivially satisfied so $\emptyset$ is accepted as an element of $\mathfrak{H}$.

$A = \{a\}$

The only possible value for x is $x = a$. Checking the definition,

$$cl(A - x) = cl(\{a\} - \{a\}) = cl(\emptyset) = \{b\}$$

and $x = a \notin \{b\}$. Thus, $\{a\}$ is an element of $\mathfrak{H}$.

$A = \{b\}$

The only possible value for x is $x = b$. Checking the definition,

$$cl(A - x) = cl(\{b\} - \{b\}) = cl(\emptyset) = \{b\}$$

and $x = b \in \{b\}$. Thus, $\{b\}$ is not an element of $\mathfrak{H}$.

$A = \{a, b\}$

There are two possible values for x. We check each separately.

First, with $x = a$,

$$cl(A - x) = cl(\{a, b\} - \{a\}) = cl(\{b\}) = \{b\}$$

and $x = a \notin \{b\}$.

Next, setting $x = b$,

$$cl(A - x) = cl(\{a, b\} - \{b\}) = cl(\{a\}) = \{a, b\}$$

and $x = b \in \{a, b\}$.

But for A to be a member of $\mathfrak{H}$, we need $x \notin cl(A - x)$ for every value of x. Since $x = b$ does not satisfy the definition, $A = \{a, b\}$ is not a member of $\mathfrak{H}$.

3.3.12 PROBLEMS

1. Let $M = (S, B)$ be the matroid with
$$S = \{a, b, c, d, e\}$$
$$B = \{\{a, b\}, \{a, c\}, \{a, d\}, \{b, c\}, \{b, d\}, \{c, d\}\}$$
Find the closure of

    1. $\{a, b, c\}$

    2. $\{a, b, e\}$

    3. $\{a, e\}$

    4. $\{c, d\}$

    5. $\{b\}$

    6. $\{e\}$

2. Let $M = (S, B)$ be the matroid with

$$S = \{a, b, c, d\}$$
$$B = \{\{a, b, d\}, \{a, c, d\}, \{b, c, d\}\}$$

Find the closure of

    1. $\{a, b, c\}$

    2. $\{b, c, d\}$

    3. $\{b, c\}$

    4. $\{a, d\}$

    5. $\{b\}$

    6. $\{d\}$

3. Find the flats of the matroid in problem 1.

4. Find the flats of the matroid in problem 2.

Chapter 4: Matroid Operations

This chapter examines the fundamental operations on matroids. These operations create new matroids by acting on one or more other matroids. For example, the unary operations of deletion and contraction create a new matroid by operating on one of the basis elements of the original matroid. In addition, we define analogous concepts for matroid sums and products in terms of matroid operations. These operators, along with the matroid dual and minors extend the general concepts of matrix algebra to the theory of matroids.

4.1   Deletion

We can create new matroids from existing matroids by the process of deletion. If M is a matroid on a set (S, T), deletion is the process of removing some elements from S, finding the subsets of T that do not contain the removed elements, and creating a matroid from the result. Essentially, we create a new matroid from the elements of S and T that do not contain the removed elements.

4.1.1   DEFINITION: DELETION

Let M be a matroid (S, B) over a set S with set of bases B. Let $R \subseteq S$ be a subset of S, and let A be the set of maximal subsets of the elements of B that do not contain an element of R. The set $(S - R, A)$ is the deletion of R from M. The new system N is designated

$$N = M \backslash R = (S - R, A).$$

We will show that the new system N is also a matroid. The process of deletion presents one way we can create a new matroid from an existing matroid. The new matroid is over the set $S - R$ which is smaller than the original matroid ground set S.

In the above definition, we chose the basis definition of a matroid as our starting point. However, we could use any of the other definitions and produce a similar result. In general, the deletion process is carried out by removing element from the ground set and creating a smaller set of independent sets.

4.1.2 EXAMPLE

Given the matroid $M = (S, B)$ with

$$S = \{a, b, c, d\}$$
$$B = \{\{a, b\}, \{a, c\}, \{a, d\}, \{b, c\}, \{b, d\}, \{c, d\}\}$$

find the deletion $N = M \backslash R$ where $R = \{c, d\}$.

First, find the new ground set

$$S - R = \{a, b\}.$$

Examine each basis element and find the maximal subset of each element that does not contain an element of R. These are the candidate sets for the basis of N:

$$B_1 = \{a, b\} \rightarrow A_1 = \{a, b\}$$

$$B_2 = \{a, c\} \rightarrow A_2 = \{a\}$$

$$B_3 = \{a, d\} \rightarrow A_3 = \{a\}$$

$$B_4 = \{b, c\} \rightarrow A_4 = \{b\}$$

$$B_5 = \{b, d\} \rightarrow A_5 = \{b\}$$

$$B_6 = \{c, d\} \rightarrow A_6 = \{\emptyset\}$$

All of the elements on this list are subsets of the first element $\{a, b\}$. We take this as the new basis $A = \{\{a, b\}\}$. The new system is $N = M \backslash R = (S - R, A)$.

The new system is also a matroid. **BM-1** is satisfied because $A \neq \emptyset$, and **BM-2** is trivially satisfied because there is only one element in A.

4.1.3 EXAMPLE

Given the matroid $M = (S, B)$ with

$$S = \{a, b, c, d\}$$
$$B = \{\{a, b\}, \{a, c\}, \{a, d\}, \{b, c\}, \{b, d\}, \{c, d\}\}$$

find the deletion $N = M \backslash R$ where $R = \{c\}$.

Find the new ground set

$$S - R = \{a, b, d\}.$$

Again, examine each basis element and find the maximal subset of each element that does not contain an element of R. These are the candidate sets for the basis of N:

$$B_1 = \{a, b\} \rightarrow A_1 = \{a, b\}$$

$$B_2 = \{a, c\} \rightarrow A_2 = \{a\}$$

$$B_3 = \{a, d\} \rightarrow A_3 = \{a, d\}$$

$$B_4 = \{b, c\} \rightarrow A_4 = \{b\}$$

$$B_5 = \{b, d\} \rightarrow A_5 = \{b, d\}$$

$$B_6 = \{c, d\} \rightarrow A_6 = \{d\}$$

The new basis will be the maximal elements from this list:

$$A = \{\{a, b\}, \{a, d\}, \{b, d\}\}.$$

The new system is also a matroid. **BM-1** is satisfied because $A \neq \emptyset$. We can check that **BM-2** is satisfied by examining each pair of distinct elements in A:

A_1, A_2:

$$A_1 - A_2 = \{b\} \quad A_2 - A_1 = \{c\} \quad (A_1 - b) \cup c = \{a, c\} = A_2$$

A_2, A_1:

$$A_2 - A_1 = \{c\} \quad A_1 - A_2 = \{b\} \quad (A_2 - c) \cup b = \{a, b\} = A_1$$

A_1, A_3:

$$A_1 - A_3 = \{a\} \quad A_3 - A_1 = \{c\} \quad (A_1 - a) \cup c = \{b,c\} = A_3$$

A_3, A_1:

$$A_3 - A_1 = \{c\} \quad A_1 - A_3 = \{a\} \quad (A_3 - c) \cup a = \{a,b\} = A_1$$

A_2, A_3:

$$A_2 - A_3 = \{a\} \quad A_3 - A_2 = \{b\} \quad (A_2 - a) \cup b = \{b,c\} = A_3$$

A_3, A_2:

$$A_3 - A_2 = \{b\} \quad A_2 - A_3 = \{a\} \quad (A_3 - b) \cup a = \{a,c\} = A_2$$

The system $N = M \backslash R = (S - R, A)$ is a matroid.

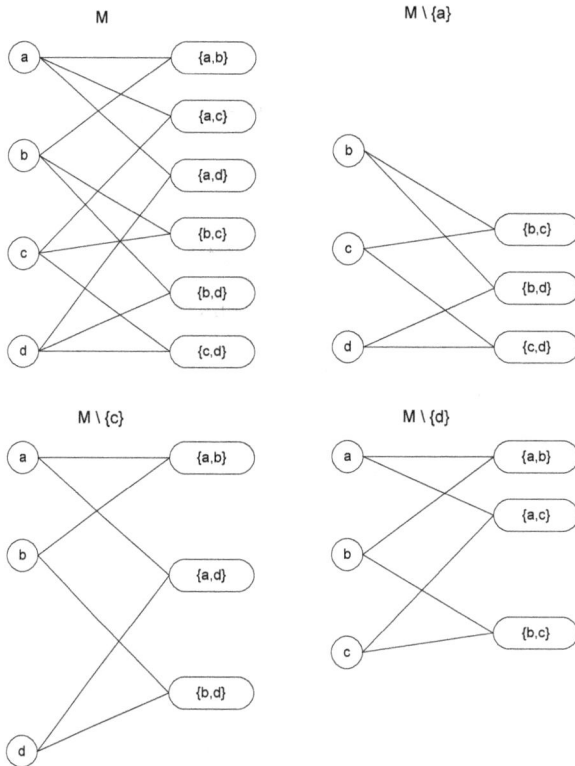

Figure 16: Deletion from a matroid. Items on the left represent the elements of the ground set, while those on the right are elements of the basis. From the original matroid M, remove the deleted element and all basis elements that contain the deleted element.

4.1.4 THEOREM

Let M be a matroid on S with set of bases B. Let N be the deletion of $R \subseteq S$; $N = M \backslash R = (S - R, A)$, where A is the set of maximal subsets of the elements of B that do not contain an element of R. Then N is a matroid.

Proof. If M is a matroid over the set S with set of bases B, then the independent set T of M is the collection of all subsets of the elements of B. Let $I \subseteq T$ be the maximal subset of T such that no element of I is an element of R.

For the moment, assume $(S - R, I)$ is a matroid. The set of bases of this matroid are the maximal independent sets of I. I is formed by finding the elements of T that do not contain an element of R.

However, the set T is the set of all subsets of the elements of B. The maximal sets of I are the maximal elements of T that do not contain an element of R. But since the elements of T are the subsets of B, the maximal sets of I are the maximal subsets of the elements of B that do not contain an element of R.

Consequently, if $(S - R, I)$, then the basis of the matroid, A, is the maximal subsets of the elements of B that do not contain an element of R.

We only need to prove that the system $(S - R, I)$ is a matroid. The system $(S - R, I)$ is a matroid because it meets the three requirements **SM-1**, **SM-2**, and **SM-3**:

SM-1 $\emptyset \in I$

We know $\emptyset \in T$ because (S, T) is a matroid. Since I is a maximal subset of T, at the very least, $\emptyset \in I$.

SM-2 If $i \in I$ and $\bar{i} \subseteq i$, then $\bar{i} \in I$

If $i \in I$ then no element of i is in R. This will also be true for every subset $\bar{i} \subseteq i$. But if $i \in I$, then $i \in T$. We know that every subset of an element in T is also in T, so $\bar{\imath} \subseteq T$. Since I is the maximal subset of T whose elements do not contain an element of R, $\bar{\imath} \subseteq I$.

SM-3   Let $i_1, i_2 \in I$ where $|i_1| < |i_2|$, then $\exists\, d \in i_2 - i_1$ such that $i_1 \cup d \in I$

If $i_1, i_2 \in I$ then $i_1, i_2 \in T$. For every pair of elements $i_1, i_2 \in T, \exists\, d \in i_2 - i_1$ such that $i_1 \cup d \in T$. However, since $d \in i_2 - i_1, d \notin R$. But d cannot be an element of R because neither i_1 nor i_2 have any elements from R. Thus, $(i_1 \cup d) \cap R = \emptyset$. So the set $i_1 \cup d$ does not have any elements from R either. But since $i_1 \cup d \in T$, and since this set does not have any elements form R, this set must be in I.

The system $(S - R, I)$. The basis of this matroid is the set A specified as the maximal subsets of the elements of B that do not contain an element of R. Thus, $N = M\backslash R = (S - R, A)$ is a matroid. ∎

4.1.5   THEOREM

Let M be a matroid (S, B) over a set S. Let $A, B \subseteq S$ be subsets of S such that $A \cap B = \emptyset$. Then

$$(M\backslash A)\backslash B = (M\backslash B)\backslash A.$$

Proof. Let I be the independent set of M. Let $X = (M\backslash A)\backslash B$ be the matroid (S_X, I_X) with ground set S_X and independent set I_X. Similarly, let $Y = (M\backslash B)\backslash A$ be the matroid (S_Y, I_Y) with ground set S_Y and independent set I_Y.

First, we know that $S_X = S - A - B$ as the process of deletion removes the elements of the deleted set from the ground set. Moreover, $S_Y = S - B - A$. But the order of removing the elements A and B from S is immaterial. So , $S_Y = S - B - A = S - A - B = S_X$.

The set I_X was created by starting with the set I, removing all elements $e \in I$ such that $e \cap A \neq \emptyset$, then removing from this set all elements e such that $e \cap B \neq \emptyset$. For every $p \in I_X$, we know $p \cap A = \emptyset$ and $p \cap B = \emptyset$.

We construct I_Y in two steps. First we remove from the set I all elements $e \in I$ such that $e \cap B \neq \emptyset$. Second, we remove from this set all elements e such that $e \cap A \neq \emptyset$. Now, for $p \in I_X$, the set p will survive the first step, because $p \cap B = \emptyset$. Furthermore, p will survive the second step because $p \cap A = \emptyset$.

Thus, if $p \in I_X$, then $p \in I_Y$. Since this is true for every element in I_X, we have $I_X \subseteq I_Y$. However, we can carry out the same argument starting with I_Y. The result is $I_Y \subseteq I_X$.

Since both $I_X \subseteq I_Y$ and $I_Y \subseteq I_X$, it must be that $I_X = I_Y$. Therefore, the two matroids X and Y have the same ground set and the same set of independent sets. This means that $X = Y$ or,

$$(M\backslash A)\backslash B = X = Y = (M\backslash B)\backslash A. \blacksquare$$

4.1.6 THEOREM

Let M be a matroid (S, B) over a set S. Let $A, B \subseteq S$ be subsets of S such that $A \cap B = \emptyset$. Then

$$(M\backslash A)\backslash B = M\backslash(A \cup B).$$

Proof. Similar to the previous proof, we see immediately that the ground sets for the matroid $(M\backslash A)\backslash B$ is the same as the ground set for $M\backslash(A \cup B)$.

Furthermore, the independents are the same as well because the independent set of $(M\backslash A)\backslash B$ is constructed from some I, removing all elements $e \in I$ such that $e \cap A \neq \emptyset$, then removing all e such that $e \cap B \neq \emptyset$. This is equivevelentt to the single step of removing all elements $e \in I$ such that $e \cap A \neq \emptyset$ or $e \cap B \neq \emptyset$, which is the same as removing all elements such that $e \cap (A \cup B) \neq \emptyset. \blacksquare$

4.1.7 PROPERTIES

We state two properties of deletion that will be useful in later sections. In these expression, M is a matroid with ground set S, r is the rank function on M, r is the rank function for the dual, and $X \subseteq S$.

DP-1 $r_{M\backslash e}(X) = r_M(X)$

DP-2 $(M\backslash A)\backslash B = (M\backslash B)\backslash A = M\backslash(A \cup B) \blacksquare$

4.1.8   PROBLEMS

1. Let $M = (S, B)$ be the matroid
$$S = \{a, b, c, d\}$$
$$B = \{\{a, c\}, \{a, d\}, \{b, c\}, \{b, d\}\}$$
Find the deletions
    a. $M \backslash \{a\}$
    b. $M \backslash \{a, b\}$
    c. $M \backslash \{a, c\}$
    d. $M \backslash \{c, d\}$

2. Let $M = (S, B)$ be the matroid
$$S = \{a, b, c, d, e\}$$
$$B = \{\{a, c, d\}, \{a, c, e\}, \{a, d, e\}, \{b, c, d\}, \{b, c, e\}, \{b, d, e\}, \{c, d, e\}\}$$
Find the deletions
    a. $M \backslash \{e\}$
    b. $M \backslash \{d, e\}$
    c. $M \backslash \{c, d, e\}$
    d. $M \backslash \{b, c, d, e\}$

3. Let $M = (S, B)$ be the matroid
$$S = \{a, b, c\}$$
$$B = \{\{a, c\}, \{b, c\}\}$$
Find the deletions
    a. $M \backslash \{a\}$
    b. $M \backslash \{b\}$
    c. $M \backslash \{c\}$
    d. $M \backslash \{a, b\}$
    e. $M \backslash \{a, c\}$
    f. $M \backslash \{b, c\}$

4.2   Dual

The notion of duals in matroids is similar to duals in graphs. This section defines the dual for a matroid and provides some basic theorems.

4.2.1   DEFINITION: MATROID DUAL

> Let M be a matroid (S, B) over a set S with set of bases B. The dual of M is the matroid with bases set $B^* = \{S - B_i \mid B_i \in B\}$.

In general, if X is a matroid concept, the equivalent dual concept is designated by X^*.

4.2.2 EXAMPLE

Let M be a matroid with

$$S = \{a, b, c\}$$

$$B = \{\{a, b\}, \{a, c\}, \{b, c\}\}$$

Find the dual of M.

The set of bases of the dual is the complement of the set of bases of M in S:

$$B_1^* = S - B_1 = \{a, b, c\} - \{a, b\} = \{c\}$$

$$B_2^* = S - B_2 = \{a, b, c\} - \{a, c\} = \{b\}$$

$$B_3^* = S - B_3 = \{a, b, c\} - \{b, c\} = \{a\}$$

The set of bases of M^* is $B^* = \{\{a\}, \{b\}, \{c\}\}$.

4.2.3 EXAMPLE

Let M be a matroid with

$$S = \{a, b, c\}$$

$$B = \{\{a\}, \{b\}, \{c\}\}$$

Find the dual of M.

The set of bases of the dual is the complement of the set of bases of M in S:

$$B_1^* = S - B_1 = \{a, b, c\} - \{a\} = \{b, c\}$$

$$B_2^* = S - B_2 = \{a, b, c\} - \{b\} = \{a, c\}$$

$$B_3^* = S - B_3 = \{a, b, c\} - \{c\} = \{a, b\}$$

The set of bases of M^* is $B^* = \{\{a,b\},\{a,c\},\{b,c\}\}$.

4.2.4   THEOREM

Show that

$$(M^*)^* = M.$$

Proof. Begin with the set S and set of bases B. The set of bases of the dual is $B_i^* = S - B_i$. The dual to this is has set of bases

$$(B_i^*) = S - B_i^* = S - (S - B_i) = B_i.$$

So the set of bases of $(M^*)^*$ is the same as the set of bases of M, and since both have set S, $(M^*)^* = M$. ∎

4.2.5   THEOREM

Let M be a matroid on a set S and M^* be the dual of M. If the rank of M is r and the rank of M^* is r^*, show that

$$r + r^* = |S|.$$

Proof. If r is the rank of M, then the dimension of every basis of M is r. To form the dual bases, we remove elements in the basis from S. Because every element in a basis must be in S, we know that the dimension of every basis of the dual is $|S| - r$. But this is the rank of the dual matroid, so $r^* = |S| - r$, or $r + r^* = |S|$. ∎

4.2.6   EXAMPLE

Let M be a matroid with

$$S = \{a,b,c,d\}$$

$$B = \{\{a,b\}\}$$

Find the dual of M.

The set of bases of the dual is the complement of the set of bases of M in S:

$$B_1^* = S - B_1 = \{a,b,c,d\} - \{a,b\} = \{c,d\}$$

The set of bases of M^* is $B^* = \{\{c, d\}\}$.

4.2.7   EXAMPLE

Let M be a matroid with

$$S = \{a, b, c, d\}$$

$$B = \{\{a, b, c\}, \{a, b, d\}\}$$

Find the dual of M.

The set of bases of the dual is the complement of the set of bases of M in S:

$$B_1^* = S - B_1 = \{a, b, c, d\} - \{a, b, c\} = \{d\}$$

$$B_2^* = S - B_2 = \{a, b, c, d\} - \{a, b, d\} = \{c\}$$

The set of bases of M^* is $B^* = \{\{c\}, \{d\}\}$.

4.2.8   EXAMPLE

Let M be a matroid with

$$S = \{a, b, c, d\}$$

$$B = \{\{c\}, \{d\}\}$$

Find the dual of M.

The set of bases of the dual is the complement of the set of bases of M in S:

$$B_1^* = S - B_1 = \{a, b, c, d\} - \{c\} = \{a, b, d\}$$

$$B_2^* = S - B_2 = \{a, b, c, d\} - \{d\} = \{a, b, c\}$$

The set of bases of M^* is $B^* = \{\{a, b, c\}, \{a, b, d\}\}$.

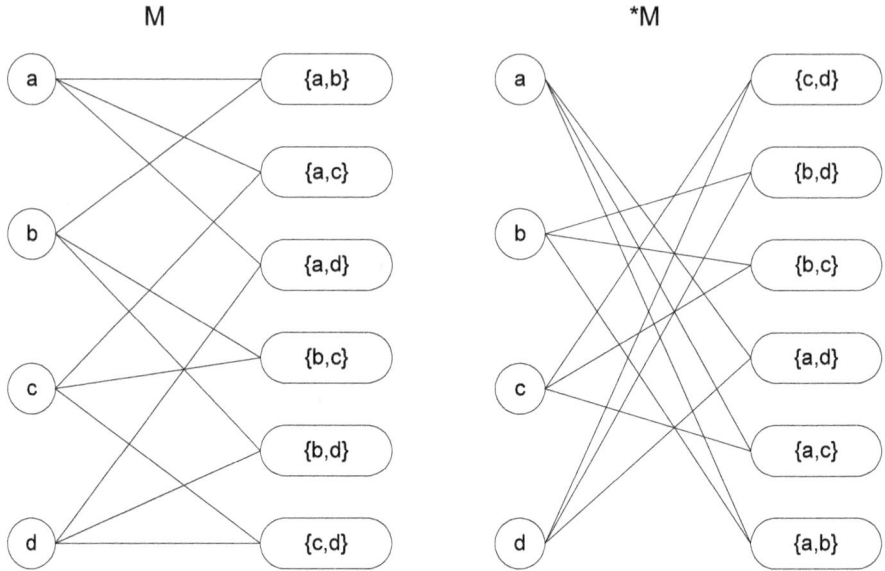

Figure 17: Dual of a matroid.

4.2.9   PROPERTIES

We state some properties of the rank of the dual that will be useful in later sections. In these expressions, M is a matroid with ground set S, r is the rank function on M, r^* is the rank function for the dual, and $X \subseteq S$.

DR-1 $r(S) + r^*(S) = |S|$

DR-2 $r^*(X) = |X| + r(S - X) - r(S)$

4.2.10   PROBLEMS

4. Let $M = (S, B)$ be the matroid
$$S = \{a, b, c, d\}$$
$$B = \{\{a, d\}, \{b, d\}, \{c, d\}\}$$
Find the dual M^*.

5. Let $M = (S, B)$ be the matroid
$$S = \{a, b, c, d\}$$
$$B = \{\{a, b, c\}, \{a, b, d\}, \{a, c, d\}, \{b, c, d\}\}$$
Find the dual M^*.

4.3 Contraction

Another method to create new matroids is by contraction. Contraction results from a combination of the operations of deletion and dual.

4.3.1 DEFINITION: CONTRACTION

Let M be a matroid over a set S. The contraction of R from M is

$$M/R = (M^*\backslash R)^*.$$

To get the contraction of R from M, we first find the dual of M, M^*. Then we delete R from M^*. Finally, we take the dual of the resulting matroid.

4.3.2 THEOREM

Let M be a matroid and let N be the contraction of R from M. N is a matroid.

Proof. M^* is a matroid because the dual of a matroid is also a matroid. Furthermore, $M^*\backslash R$ is a matroid because the deletion of R from M is a matroid. Finally, because the dual of a matroid is also a matroid, we know $N = (M^*\backslash R)^*$ must be a matroid. ∎

4.3.3 EXAMPLE

Given the matroid $M = (S, B)$ with

$$S = \{a, b, c, d\}$$
$$B = \{\{a, b\}, \{a, c\}, \{a, d\}, \{b, c\}, \{b, d\}, \{c, d\}\}$$

find the contraction $N = M/R$ where $R = \{c\}$.

First, find the dual of M. The set of bases of the M^* is

$$B_1^* = S - B_1 = \{a, b, c, d\} - \{a, b\} = \{c, d\}$$

$$B_2^* = S - B_2 = \{a, b, c, d\} - \{a, c\} = \{b, d\}$$

$$B_3^* = S - B_3 = \{a, b, c, d\} - \{a, d\} = \{b, c\}$$

$$B_4^* = S - B_4 = \{a, b, c, d\} - \{b, c\} = \{a, d\}$$

$$B_5^* = S - B_5 = \{a, b, c, d\} - \{b, d\} = \{a, c\}$$

$$B_6^* = S - B_6 = \{a, b, c, d\} - \{c, d\} = \{a, b\}$$

Next, delete R from M^*. The new ground set is

$$S - R = \{a, b, d\},$$

and the candidate sets for the new bases are:

$$B_1^* = \{c, d\} \rightarrow A_1 = \{d\}$$

$$B_2^* = \{b, d\} \rightarrow A_2 = \{b, d\}$$

$$B_3^* = \{b, c\} \rightarrow A_3 = \{b\}$$

$$B_4^* = \{a, d\} \rightarrow A_4 = \{a, d\}$$

$$B_5^* = \{a, c\} \rightarrow A_5 = \{a\}$$

$$B_6^* = \{a, b\} \rightarrow A_6 = \{a, b\}$$

The new basis will be the maximal elements from this list:

$$A = \{\{a, b\}, \{a, d\}, \{b, d\}\}.$$

Finally, take the dual of the matroid $(S - R, A)$. The set of bases for the dual is

$$Z_1 = (S - R) - B_1^* = \{a, b, d\} - \{a, b\} = \{d\}$$

$$Z_2 = (S - R) - B_2^* = \{a, b, d\} - \{a, d\} = \{b\}$$

$$Z_3 = (S - R) - B_3^* = \{a, b, d\} - \{b, d\} = \{a\}$$

So the contraction of R from M is the matroid over the set $S - R$ with bases set Z:

$$N = (M^* \backslash R)^* = (S - R, Z).$$

4.3.4   EXAMPLE

Given the matroid $M = (S, B)$ with

$$S = \{a, b, c\}$$
$$B = \{\{a\}, \{b\}, \{c\}\}$$

find the contraction $N = M/R$ where $R = \{c\}$.

Again, we go through three steps. First, find the dual M^*. Second, find the deletion of R from M^*. Third, take the dual of the result.

Step 1: Find M^*

$$B_1^* = S - B_1 = \{a, b, c\} - \{a\} = \{b, c\}$$

$$B_2^* = S - B_1 = \{a, b, c\} - \{b\} = \{a, c\}$$

$$B_3^* = S - B_1 = \{a, b, c\} - \{c\} = \{a, b\}$$

Step 2: Find $M^*\backslash R$

$$S - R = \{a, b\},$$

The candidate sets for the new bases are:

$$B_1^* = \{b, c\} \rightarrow A_1 = \{b\}$$

$$B_2^* = \{a, c\} \rightarrow A_2 = \{a\}$$

$$B_3^* = \{a, b\} \rightarrow A_3 = \{a, b\}$$

Since the new bases set is the set of maximal sets, we have $A = \{\{a, b\}\}$.

Step 3: Find $N = (M^*\backslash R)^*$

$$Z_1 = (S - R) - A_1 = \{a, b\} - \{a, b\} = \emptyset$$

Putting everything together, the contraction of R from M is the matroid $N = (M^*\backslash R)^*$ which is the matroid over the set $S - R$ with set of bases $Z = \{\emptyset\}$.

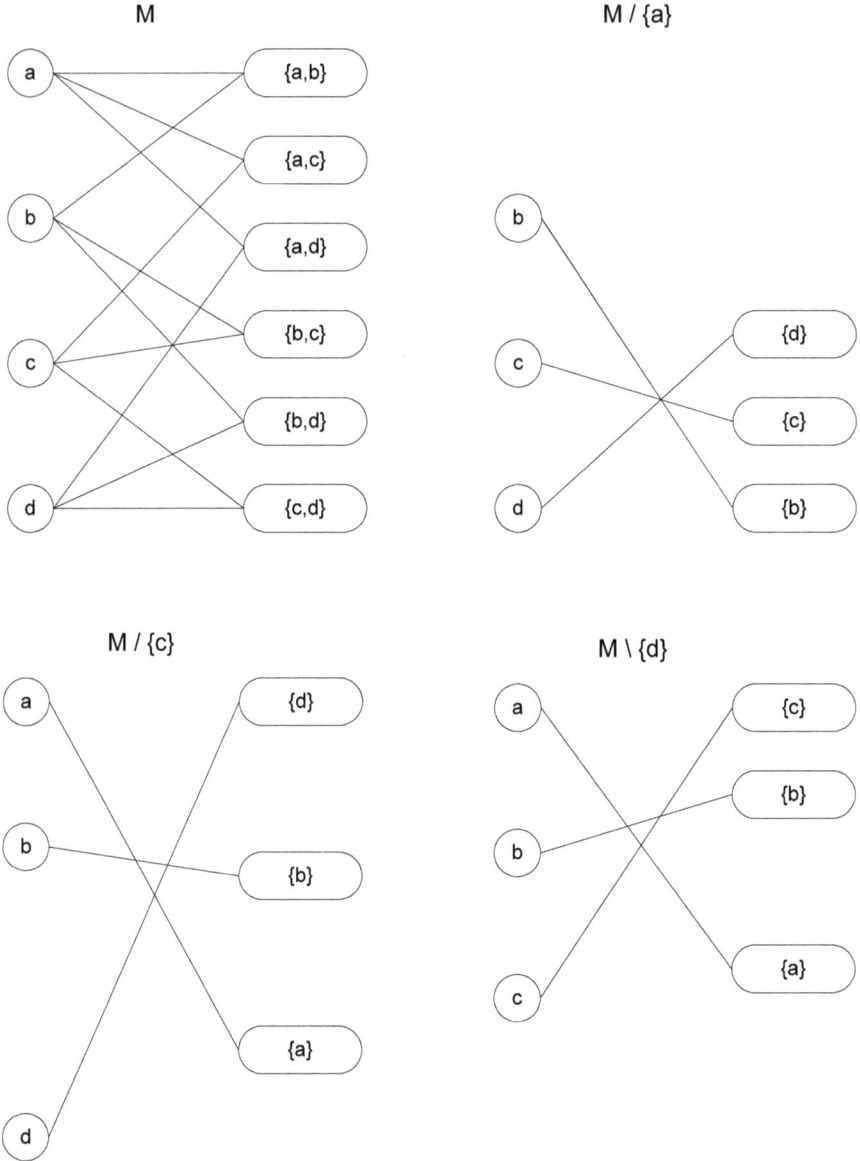

Figure 18: Contraction from a matroid.

4.3.5 THEOREM

Let M be a matroid (S, B) over a set S. Let $A, B \subseteq S$ be subsets of S such that $A \cap B = \emptyset$. Then

$$(M/A)/B = (M/B)/A.$$

Proof. We know

$$X/Y \equiv (X^* \backslash Y)^*.$$

Now, $M/A \equiv (M^* \backslash A)^*$. If we set $X = (M^* \backslash A)^*$ and $Y = B$,

$$X/Y = (X^* \backslash Y)^* = ((M^* \backslash A)^{**} \backslash B)^* = ((M^* \backslash A) \backslash B)^*.$$

But from theorem 4.3.5 we know $(S \backslash A) \backslash B = (S \backslash B) \backslash A$. Using this we have,

$$(M^* \backslash A) \backslash B = (M^* \backslash B) \backslash A.$$

Substituting this into the previous result,

$$((M^* \backslash A) \backslash B)^* = ((M^* \backslash B) \backslash A)^* = ((M^* \backslash B)^{**} \backslash A)^* = (M/B)/A.$$

Finally we have, $(M/A)/B = (M/B)/A.$ ■

4.3.6   THEOREM

Let M be a matroid (S, B) over a set S. Let $A, B \subseteq S$ be subsets of S such that $A \cap B = \emptyset$. Then

$$(M/A)/B = M/(A \cup B).$$

Proof. From the previous proof we have

$$(M/A)/B = ((M^* \backslash A) \backslash B)^*.$$

But,

$$(M^* \backslash A) \backslash B = M^* \backslash (A \cup B).$$

From this we know,

$$(M/A)/B = ((M/A)/B)^* = (M^* \backslash (A \cup B))^* = M/(A \cup B). ■$$

4.3.7   PROPERTIES

We state some properties of the contraction that will be useful in later sections. In this expression, M is a matroid with ground set S, r is the rank function on M, r is the rank function for the dual, and $X \subseteq S$.

CP-1 $r_{M/T}(X) = r_M(X \cup T) - r_M(T)$

CP-2 $M - e = M/e$ if and only if $r(X) + r(S - X) = r(M)$

From this, $M - e = M/e$ if and only if e is a loop or an isthmus.

CP-3 $(M/A)/B = (M/B)/A = M/(A \cup B)$

CP-4 $(M/A)\backslash B = (M\backslash B)/A$

4.3.8 PROBLEMS

6. Let $M = (S, B)$ be the matroid
$$S = \{a, b, c\}$$
$$B = \{\{a, c\}, \{b, c\}\}$$
Find the contractions
   a. $M/\{a\}$
   b. $M/\{b\}$
   c. $M/\{c\}$
   d. $M/\{a, b\}$
   e. $M/\{a, c\}$
   f. $M/\{b, c\}$

7. Let $M = (S, B)$ be the matroid
$$S = \{a, b, c\}$$
$$B = \{\{a\}, \{b\}, \{c\}\}$$
Find the contractions
   a. $M/\{a\}$
   b. $M/\{b\}$
   c. $M/\{c\}$
   d. $M/\{a, b\}$
   e. $M/\{a, c\}$
   f. $M/\{b, c\}$

8. Let $M = (S, B)$ be the matroid
$$S = \{a, b, c, d\}$$
$$B = \{\{a, d\}, \{b, d\}, \{c, d\}\}$$
Find the contractions
   a. $M/\{c\}$
   b. $M/\{d\}$

4.4 Minors

Matroid minors provide insight to the structure of the matroid. Minors are constructed from the operations of contraction and deletion.

4.4.1 DEFINITION: MINORS

> Let M be a matroid over a set S, and let X and Y be disjoint subsets of S. A minor of X is
>
> $$M \backslash X / Y = M / Y \backslash X$$

4.4.2 EXAMPLE

Given the matroid $M = (S, B)$ with

$$S = \{a, b, c\}$$
$$B = \{\{a, b\}\}$$

find the minor using $X = \{a\}$ and $Y = \{b\}$.

First, compute $N = M \backslash X$:

The new ground set is

$$S - X = \{b, c\}.$$

The candidate sets for the basis are:

$$B_1 = \{b\} \rightarrow A_1 = \{b\},$$

so the set of bases is $A = \{\{b\}\}$, and $N = (S - X, A)$.

Second, compute N / Y:

Find the dual of N. The bases are:

$$A_1^* = N - A_1 = \{b, c\} - \{c\} = \{c\}.$$

The dual is then the matroid $N^* = (S - X, A^*)$

Now delete Y from N^*.

The new ground set is

$$(S - X) - Y = \{c\}.$$

The candidate sets for the basis are:

$$C_1 = \{c\} \rightarrow D_1 = \{c\},$$

so the set of bases is $D = \{\{c\}\}$, and $N^*/Y = (S - X - Y, D)$.

Third, take the dual of N^*/Y. The set of bases for this dual is

$$Z_1 = (S - X - Y) - D_1 = \{c\} - \{c\} = \emptyset.$$

The minor is

$$M \backslash X / Y = (S - X - Y, Z) = (\{c\}, \{\emptyset\}).$$

4.4.3   EXAMPLE

Given the matroid $M = (S, B)$ with

$$S = \{a, b, c, d\}$$
$$B = \{\{a, b, c\}, \{a, b, d\}\}$$

find the minor using $X = \{a, b\}$ and $Y = \{c\}$.

First, compute $N = M \backslash X$:

The new ground set is

$$S - X = \{c, d\}.$$

The candidate sets for bases are:

$$B_1 = \{a, b, c\} \rightarrow A_1 = \{c\},$$

$$B_2 = \{a, b, d\} \rightarrow A_2 = \{d\}$$

so the set of bases is $A = \{\{c\}, \{d\}\}$, and $N = (S - X, A)$.

Second, compute N/Y:

Find the dual of N. The bases are:

$$A_1^* = N - A_1 = \{c, d\} - \{c\} = \{d\}.$$

$$A_2^* = N - A_1 = \{c, d\} - \{d\} = \{c\}.$$

The dual is then the matroid $N^* = (S - X, A^*)$

Now delete Y from N^*.

The new ground set is

$$(S - X) - Y = \{d\}.$$

The candidate sets for the basis are:

$$C_1 = \{d\} \rightarrow D_1 = \{d\},$$

$$C_2 = \{c\} \rightarrow D_2 = \{\emptyset\},$$

so the set of bases is $D = \{\{d\}\}$, and $N^*/Y = (S - X - Y, D)$.

Third, take the dual of N^*/Y. The set of bases for this dual is

$$Z_1 = (S - X - Y) - D_1 = \{d\} - \{s\} = \{\emptyset\}.$$

The minor is

$$M\backslash X/Y = (S - X - Y, Z) = (\{d\}, \{\{\emptyset\}\}).$$

4.4.4 THEOREM

Let M be a matroid over a set S, and let X and Y be disjoint subsets of S. The minor $N = M\backslash X/Y$ is a matroid.

Proof. We know that the operations of contraction and deletion both result in matroids. Thus, $M\backslash X$ is a matroid, and taking the contraction of $M\backslash X$ results again in a matroid. So the minor $N = M\backslash X/Y$ is a matroid. ∎

4.5 Sums

Two matroids may be added together by joining their ground sets and every combination of independent sets. In this section we will define the Nash-Williams matroid sum and provide some examples.

4.5.1 DEFINITION: NASH-WILLIAMS SUM

> Let M_1 be the matroid (S_1, T_1) and M_2 the matroid (S_2, T_2). The Nash-Williams sum $M_3 = M_1 \oplus M_2$ is the system (S_3, T_3) and is a matroid where
>
> **SUM-1** $S_3 = S_1 \cup S_2$
>
> **SUM-2** $T_3 = \{t_1 \cup t_2, t_1 \in T_1, t_2 \in T_2\}$

We can compute the sum from the set of bases as well. Let the set of bases of M_1 be designated as B_1 and the set of bases of M_2 designated as B_2. The set of bases for M_3 is the maximal set

$$B_3 = [\{b_1 \cup b_2, b_1 \in B_1, b_2 \in B_2\}].$$

4.5.2 EXAMPLE

Let M_1 and M_2 be the matroids

$$M_1 = \begin{cases} S_1 = \{a, b, c, d\} \\ B_1 = \{\{a, b\}\} \end{cases} \quad M_2 = \begin{cases} S_2 = \{a, e\} \\ B_2 = \{\{a\}, \{e\}\} \end{cases}.$$

Find the sum $M_3 = M_1 \oplus M_2$.

The ground set for the new matroid is

$$S_3 = S_1 \cup S_2 = \{a, b, c, d\} \cup \{a, e\} = \{a, b, c, d, e\}.$$

The candidates for the bases are

$$\bar{B}_1 = \{a, b\} \cup \{a\} = \{a, b\} \quad \bar{B}_2 = \{a, b\} \cup \{e\} = \{a, b, e\}.$$

We see that $\bar{B}_1 \subset \bar{B}_2$, so only $\bar{B}_2$ is a basis for M_3. Thus

$$M_3 = \begin{cases} S_3 = \{a, b, c, d, e\} \\ B_3 = \{\{a, b, e\}\} \end{cases}.$$

4.5.3 EXAMPLE

Let M_1 and M_2 be the matroids

$$M_1 = \begin{cases} S_1 = \{a,b,c,d\} \\ B_1 = \{\{a,b,c\},\{a,b,d\}\} \end{cases} \quad M_2 = \begin{cases} S_2 = \{a,b,c\} \\ B_2 = \{\{a\},\{b\},\{c\}\} \end{cases}$$

Find the sum $M_3 = M_1 \oplus M_2$.

The ground set for the new matroid is

$$S_3 = S_1 \cup S_2 = \{a,b,c,d\} \cup \{a,b,c\} = \{a,b,c,d\}.$$

The candidates for the bases are

$$\bar{B}_1 = \{a,b,c\} \cup \{a\} = \{a,b,c\} \quad \bar{B}_2 = \{a,b,d\} \cup \{a\} = \{a,b,d\}$$

$$\bar{B}_3 = \{a,b,c\} \cup \{b\} = \{a,b,c\} \quad \bar{B}_4 = \{a,b,d\} \cup \{b\} = \{a,b,d\}$$

$$\bar{B}_5 = \{a,b,c\} \cup \{c\} = \{a,b,c\} \quad \bar{B}_6 = \{a,b,d\} \cup \{c\} = \{a,b,c,d\}$$

We see that all of these sets are subsets of $\bar{B}_6$. The matroid sum is

$$M_3 = \begin{cases} S_3 = \{a,b,c,d\} \\ B_3 = \{\{a,b,c,d\}\} \end{cases}$$

4.5.4 EXAMPLE

Let M_1 and M_2 be the matroids

$$M_1 = \begin{cases} S_1 = \{a,b,c,d\} \\ B_1 = \{\{a,b\},\{a,c\}\} \end{cases} \quad M_2 = \begin{cases} S_2 = \{a,b,c,d\} \\ B_2 = \{\{b\},\{c\},\{d\}\} \end{cases}$$

Find the sum $M_3 = M_1 \oplus M_2$.

The ground set for the new matroid is

$$S_3 = S_1 \cup S_2 = \{a,b,c,d\} \cup \{a,b,c\} = \{a,b,c,d\}.$$

The candidates for the bases are

$$\bar{B}_1 = \{a,b\} \cup \{b\} = \{a,b\} \quad \bar{B}_2 = \{a,c\} \cup \{b\} = \{a,b,c\}$$

$$\bar{B}_3 = \{a,b\} \cup \{c\} = \{a,b,c\} \quad \bar{B}_4 = \{a,c\} \cup \{c\} = \{a,c\}$$

$$\bar{B}_5 = \{a,b\} \cup \{d\} = \{a,b,d\} \quad \bar{B}_6 = \{a,c\} \cup \{d\} = \{a,c,d\}$$

There are three maximal sets in this group: $\{a,b,c\}, \{a,b,d\}$, and $\{a,c,d\}$. The matroid sum is

$$M_3 = \begin{cases} S_3 = \{a,b,c,d\} \\ B_3 = \{\{a,b,c\},\{a,b,d\},\{a,c,d\}\} \end{cases}$$

4.5.5   LEMMA

Let M_1 and M_2 be matroids of rank r_1 and r_2 respectively. The rank of their sum $M_3 = M_1 \oplus M_2$ satisfies

$$r(M_3) \le r_1 + r_2.$$

Proof. The rank of M_3 is the size of any of the elements in the set of bases. From the construction of the set of bases of M_3, every basis b_3 of M_3 constructed from the union of a basis b_1 of M_1 and b_2 of M_2, so

$$|b_3| \le |b_1| + |b_2|.$$

But if b_1 is a basis of M_1 then $|b_1| = r_1$, and similarly $|b_2| = r_2$. Substituting these expressions into the inequality above,

$$|b_3| \le r_1 + r_2.$$

This is true for every candidate basis b_3. But we also know that the size of any basis of M_3 has dimension $r(M_3)$. Therefore,

$$r(M_3) \le r_1 + r_2. \ \blacksquare$$

4.5.6   PROBLEMS

9. Let $M_1 = (S_1, B_1)$ be the matroid
$$S_1 = \{a,b\}$$
$$B_1 = \{\{a,b\}\}$$
and $M_2 = (S_2, B_2)$ be the matroid
$$S_2 = \{a,b\}$$
$$B_2 = \{\{a\},\{b\}\}$$
Find the sum $M_3 = M_1 \oplus M_2$.

10. Let $M_1 = (S_1, B_1)$ be the matroid

$$S_1 = \{a, b\}$$
$$B_1 = \{\{a, b\}\}$$

and $M_2 = (S_2, B_2)$ be the matroid

$$S_2 = \{a, b, c\}$$
$$B_2 = \{\{a\}, \{b\}\}$$

Find the sum $M_3 = M_1 \oplus M_2$.

11. Let $M_1 = (S_1, B_1)$ be the matroid

$$S_1 = \{a, b\}$$
$$B_1 = \{\{a, b\}\}$$

and $M_2 = (S_2, B_2)$ be the matroid

$$S_2 = \{a, b, c\}$$
$$B_2 = \{\{a\}, \{b\}, \{b\}\}$$

Find the sum $M_3 = M_1 \oplus M_2$.

12. Let $M_1 = (S_1, B_1)$ be the matroid

$$S_1 = \{a, b, c\}$$
$$B_1 = \{\{a\}, \{b\}, \{b\}\}$$

and $M_2 = (S_2, B_2)$ be the matroid

$$S_2 = \{a, b, c, d, e\}$$
$$B_2 = \{\{a, c, d, e\}, \{b, c, d, e\}\}$$

Find the sum $M_3 = M_1 \oplus M_2$.

4.6 Matroid Product

Two matroids may have the same underlying structure even though their set of bases may appear different. In this case the matroids are called isomorphic, and the different bases result from permutations of the ground set.

In enumerating matroids, it is customary to only count matroids that have a different structure rather than just a different presentation of the ground set. The matroid product defined here will be useful in obtaining new matroids by combining two other matroids.

The matroid product is a binary operation between two matroids that creates a new object that is not necessarily a matroid. However, we can identify sufficient conditions to determine when the result is a matroid.

4.6.1 DEFINITION: MATROID PRODUCT

Let M_1 be a matroid with rank r, ground set S and set of bases B_1. Let M_2 a matroid with rank r, ground set S and set of bases B_2. The matroid product is defind as the system $M_3 = M_1 \otimes M_2$ is the system (S, B_3) where

PRD-1 $B_3 = B_1 \cup B_2$

The set B_3 is the 'basis' for M_3 and every subset of B_3 is considered part of the 'independent' set of M_3.

The structure M_3 is not necessarily a matroid. In order for M_3 to be a matroid, the set B_3 must satisfy **BM-1** and **BM-2**. The next theorem provides a sufficient condition for M_3 to be a matroid.

4.6.2 THEOREM

Let $M_3 = M_1 \otimes M_2$ be a structure formed as the product between two matroids M_1 and M_2 with rank r, ground set S, and set of bases B_1 and B_2 respectively. M_3 is a matroid if and only if:

PRM-1 For every $b_1 \in B_1$ and $b_2 \in B_2$, every element $e \in b_1 - b_2$, there exists an element $\bar{e} \in b_2 - b_1$ such that $(b_1 - e) \cup \bar{e} \in B_3$.

PRM-2 For every $b_1 \in B_1$ and $b_2 \in B_2$, every element $e \in b_2 - b_1$, there exists an element $\bar{e} \in b_1 - b_2$ such that $(b_2 - e) \cup \bar{e} \in B_3$.

Proof. If M_3 is a matroid with basis B_3, then B_3 must satisfy **BM-1** and **BM-2**. Since both M_1 and M_2 are matroids, then from **BM-1** we know $B_1 \neq \emptyset$ and $B_2 \neq \emptyset$. Thus, since $B_3 = B_1 \cup B_2 \neq \emptyset$, **BM-1** is satisfied for B_3.

In order for B_3 to satisfy **BM-2**, we need to show that for every distinct $b_1, b_2 \in B_3$, and every element $e \in b_1 - b_2$, there exists an element $\bar{e} \in b_2 - b_1$ such that $(b_1 - e) \cup \bar{e} \in B_3$. If both b_1 and b_2 are members of B_1, then **BM-2** is satisfied because B_1 satisfies BM-2 since M_1 is a matroid. Similarly, if both b_1 and b_2 are members of B_2, then **BM-2** is satisfied because M_2 is a matroid.

The only cases we need to be concerned with are when $b_1 \in B_1$ and $b_2 \in B_2$, and when $b_1 \in B_2$ and $b_2 \in B_1$. Since $B_1 \cap B_2 = \emptyset$, we do not need to consider cases where $b_1 \in B_1$ and $b_1 \in B_2$.

Examining the case where $b_1 \in B_1$ and $b_2 \in B_2$, **BM-2** states for every $b_1 \in B_1$ and $b_2 \in B_2$, and every element $e \in b_1 - b_2$, there exists an element $\bar{e} \in b_2 - b_1$ such that $(b_1 - e) \cup \bar{e} \in B_3$.

Similarly, the case where $b_1 \in B_2$ and $b_2 \in B_1$, **BM-2** states for every $b_1 \in B_2$ and $b_2 \in B_1$, and every element $e \in b_1 - b_2$, there exists an element $\bar{e} \in b_2 - b_1$ such that $(b_1 - e) \cup \bar{e} \in B_3$.

Thus, if M_3 is a matroid, **PRM-1** and **PRM-2** are satisfied.

Alternatively, if **PRM-1** and **PRM-2** are satisfied then **BM-2** is satisfied in the cases where $b_1 \in B_1$ and $b_2 \in B_2$, and when $b_1 \in B_2$ and $b_2 \in B_1$. Moreover, **BM-2** is also satisfied in the case where both b_1 and b_2 are members of B_1 because B_1 satisfies **BM-2**. Similarly, **BM-2** is satisfied when both b_1 and b_2 are members of B_2 because B_2 satisfies **BM-2**. Thus, **BM-2** is satisfied.

BM-1 is also satisfied because $B_1 \neq \emptyset$, $B_2 \neq \emptyset$, and $B_3 = B_1 \cup B_2 \neq \emptyset$.

Therefore, if **PRM-1** and **PRM-2** are true then M_3 is a matroid. ∎

4.6.3 EXAMPLE

Show that $M_3 = M_1 \otimes M_2$ is a matroid where M_1 has ground set $S = \{a, b, c\}$ and basis $B_1 = \{\{a, b\}\}$ and M_2 has ground set $S = \{a, b, c\}$ and basis $B_2 = \{\{a, c\}, \{b, c\}\}$.

The matroid product is M_3 with ground set $S = \{a, b, c\}$ and 'basis' $B = \{\{a, b\}, \{a, c\}, \{b, c\}\}$. Examining **PRM-1**, there is only one basis in B_1 so $b_1 = \{a, b\}$. There are two choices for b_2: $\{a, c\}$ or $\{b, c\}$. Checking each,

$$b_1 = \{a, b\}, \; b_2 = \{a, c\}$$

$$b_1 - b_2 = \{b\} \quad b_2 - b_1 = \{c\}$$

We have $(b_1 - b) \cup c = \{a, c\} \in B$, satisfying **PRM-1**.

$$b_1 = \{a, b\}, \; b_2 = \{b, c\}$$

$$b_1 - b_2 = \{a\} \quad b_2 - b_1 = \{c\}$$

Again, we see $(b_1 - a) \cup c = \{b, c\} \in B$, satisfying **PRM-1**. In each case **PRM-1** is satisfied.

PRM-2 is handled similarly. Checking each case:

$$b_1 = \{a, b\}, \; b_2 = \{a, c\}$$

$$b_1 - b_2 = \{b\} \quad b_2 - b_1 = \{c\}$$

We have $(b_2 - c) \cup b = \{a, b\} \in B$, satisfying **PRM-2**.

$$b_1 = \{a, b\}, \; b_2 = \{b, c\}$$

$$b_1 - b_2 = \{a\} \quad b_2 - b_1 = \{c\}$$

Similar to the previous case, $(b_2 - c) \cup a = \{a, b\} \in B$, satisfying **PRM-2**.

Both **PRM-1** and **PRM-2** are satisfied, so M_3 is a matroid.

$$P = \begin{pmatrix} a & b & c \\ c & a & b \end{pmatrix}$$

$$M_3 = M_1 \otimes M_2 \implies PM_3 = (PM_1) \otimes (PM_2)$$

B_1	{a,b}		{b,c}	=	PB_1
B_2	{a,c}		{b,a}	=	PB_2
	{b,c}		{c,a}		

Figure 19: The permutation of the matroid $M_3 = M_1 \otimes M_2$ is equivalent to the matroid product $P_k M_3 = (P_k M_1) \otimes (P_k M_2)$.

4.6.4 THEOREM

Let $M_3 = M_1 \otimes M_2$ be a matroid from a matroid product. The presentation of M_3 from a permutation P_k is the matroid product $P_k M_3 = (P_k M_1) \otimes (P_k M_2)$.

Proof. The set of presentations of M_3 are the distinct elements $P_k M_3$. Applying a permutation on the set of bases of M_3 can be carried out one basis at a time, and the transformation on each basis is dependant only on the permutation, not the other bases. Applying the permutation to each basis transforms the bases of M_1 and M_2 separately. The resulting set is equivalent to the matroid product of the permutated matroids. This process is diagrammed in Figure 19. ∎

Chapter 5: Invariants

Matroid invariants are quantities computed from a matroid representation that have the same value for every presentation of a matroid. Invariants are useful when trying to determine if two different presentations have the same matroid structure.

Invariants are useful when comparing two matroids to determine if the matroids are different. If two matroids have different values for the invariant, then the matroids cannot be the same. However, just because the matroids have the same value of the invariant, this does not mean that they must be the same.

In general, two different matroids can produce the same invariant value even through the matroids are actually different. Thus, producing the same value for an invariant is a necessary condition for two matroids to be the same, but it is not a sufficient condition.

5.1   Posets

Posets are partially ordered sets. A poset is a set with a binary function mapping pairs of elements of the set and identifies if the first element is less than or equal ($\leq$) to the second element. Posets are useful to impose an ordering relation on sets.

5.1.1   DEFINITION: POSET

> Let A be a set. Let $\leq: A \times A \rightarrow \{1\}$. The pair $(A, \leq)$ forms a poset if $\leq$ satisfies the order relations:
>
> **OR-1** $x \leq x$ (antisymmetry)
>
> **OR-2**   If $x \leq y$ and $y \leq x$ then $x = y$ (reflexivity)
>
> **OR-3**   If $x \leq y$ and $y \leq z$ then $x \leq z$ (transitivity)

The $\leq$ map is a binary function on A that maps pairs of elements of A to $\{1\}$ (true). This function is does not need to be injective; we do not need to have a value for the function for every pair of elements of A.

The three properties of $\leq$ are designed to make the operator conform to the usual ordering operation for real numbers. However, unlike real numbers, not every pair of elements need be comparable.

Finally, if x and y are distinct $(x \neq y)$, we may designate the function as $<$ instead of $\leq$.

5.1.2   EXAMPLE

Let $A = \{a, b, c, d\}$ be a set with $\leq: A \times A \rightarrow \{0,1\}$ defined by

$\leq$	a	b	c	d
a	1	1		1
b	0	1		1
c			1	
d	0	0		1

Note that some pairs such as a and c are not defined under the $\leq$ map. It is not necessary that every pair of elements be comparable.

The pair $(A, \leq)$ is a poset. We can verify each of the properties of the function $\leq$.

PR-1 $x \leq x$

$$a \leq a \rightarrow 1 \quad b \leq b \rightarrow 1 \quad c \leq c \rightarrow 1 \quad d \leq d \rightarrow 1$$

PR-2   If $x \leq y$ and $y \leq x$ then $x = y$

This is obviously satisfied for the elements on the diagonal of the table. Examining the off diagonal elements, we see that there are no pairs of elements such that $x \leq y$ and $y \leq x$.

Examine the transpose of the table. At each place where there is not entry, there is also no entry in the corresponding transpose.

Similarly, where there is an off-diagonal entry, there is also an entry on the transpose, but the transpose exchanges 0 and 1. This is sufficient to support **PR-2**.

For example, from the table we see that $a \leq b$. But we also have $b \not\leq a$ from the transpose. This is consistent with **PR-2**. If the transpose had provided that $b \leq a$, then **PR-2** would be violated.

In general when we examine a table, we only need to examine the off-diagonal elements to determine if **PR-2** is satisfied. **PR-2** is satisfied if:

1. Every off-diagonal element that is not defined is also not defined in the transpose.
2. Every off-diagonal element that is defined has opposite value in the transpose.

PR-3   If $x \leq y$ and $y \leq z$ then $x \leq z$

Again, only the off diagonal elements are needed. We need to examine each triple (x, y, z) of off diagonal elements to verify that **PR-3** is satisfied.

There are only three off diagonal elements, so we only have one triple to verify, (a, b, d). We see from the table that $a \leq b$, and $b \leq d$. We also have $a \leq d$. So **PR-3** is satisfied.

5.1.3   EXAMPLE

Let $A = \{a, b, c, d\}$ be a set with $\leq: A \times A \to \{0,1\}$ defined by

$\leq$	a	b	c	d	e
a	1		1	1	1
b		1	1	1	1
c	0	0	1	1	1
d	0	0	0	1	
e	0	0	0		1

The pair $(A, \leq)$ is a poset. We can check each of the properties of the function $\leq$ to verify that $(A, \leq)$ is a poset.

PR-1 $x \leq x$

$a \leq a \to 1 \quad b \leq b \to 1 \quad c \leq c \to 1 \quad d \leq d \to 1 \quad e \leq e \to 1$

PR-2   If $x \leq y$ and $y \leq x$ then $x = y$

Similar to the previous example, there are no pairs of elements such that $x \leq y$ and $y \leq x$ because the transpose elements have opposite value, and the undefined elements are also undefined in the transpose.

PR-3   If $x \leq y$ and $y \leq z$ then $x \leq z$

There are several triples to verify. If we pick a pair of elements (x, y) where the function is defined, we only need to check all elements z such that (y, z) is defined, and verify for each that (a, z) is defined.

(a, c)

For c, there are two mappings: (c, d) and (c, e). Examining the table we see both (a, d) and (a, e) are defined.

(a, d)

For d, there is one mapping: (d, e). Examining the table we see (a, e) is defined, so **PR-3** is satisfied for (a, d).

(a, e)

For e, there are no other mappings defined, so **PR-3** is trivially satisfied. This is true for any element (x, e), so we need not check any other elements of the form (x, e) because they will all satisfy **PR-3**.

(b, c)

Again there are two mappings for c: (c, d) and (c, e). Examining the table we see both (b, d) and (b, e) are defined.

(b, d)

Similarly, d has only one mapping: (d, e). From the table (b, e) is defined.

(c, d)

Finally, d again has only one mapping: (d, e). From the table (c, e) is defined, so **PR-3** is satisfied.

We see that **PR-3** is satisfied for every triple (x, y, z). Since **PR-1**, **PR-2**, and **PR-3** are all satisfied, $\leq$ is an order relation and the pair $(A, \leq)$ is a poset.

5.1.4   DEFINITION: HASSE DIAGRAM

Let P be a poset. A Hasse diagram is a graph in which each vertex represents an element $p \in P$, and a line is drawn between distinct vertices p_1 and p_2 if $p_1 < p_2$ and there is no element p_3 such that $p_1 < p_3 < p_2$.

5.1.5   EXAMPLE

Figure 20 shows a Hasse diagram for the poset of Example 5.1.2. In addition, Figure 21 is the Hasse diagram for the poset of Example 5.1.3.

The Hasse diagrams are useful to visually represent the poset relations. The elements are ordered from bottom to top in the sense that elements appearing lower in the diagram are less than elements that are higher. But this is only the case if there the elements are connected by a sequence of lines moving upward in the diagram.

For example, in Figure 20, c is not connected to any other element. Even though c appears higher in the diagram than a, this does not mean that a is less than c because there is no upward moving path connecting a to c.

Similarly, in Figure 21, d is not comparable to e because there is no upward moving path connecting d to e. There are paths connecting d

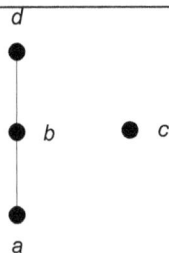

Figure 20: Hasse diagram for the poset in Example 5.1.2.

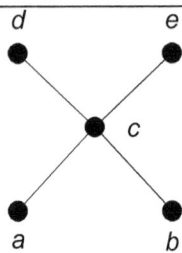

Figure 21: Hasse diagram for the poset in Example 5.1.3.

to e, but we have to first move down from d to c, then up from c to e. Two elements are only comparable if there is an upward moving path connecting them.

However, b and e of Figure 21 are comparable. Although these elements are not connected by a single line, there is an upward moving path connecting them. There is an upward connection from b to c, and an updard connection from c to e.

5.1.6 DEFINITION: COVER

Let P be a poset. Let $p_1, p_2 \in P$. If $p_1 < p_2$ and there is no $p_3 \in P$ such that $p_1 < p_3 < p_2$, then p_2 covers p_1.

From the definition of the Hasse diagram, we draw a line between two elements if one is a cover of the other.

5.2 Lattice of Flats

A lattice of flats may be associated with every matroid. Matroid flats were defined in section 3.3.6. The lattice of flats does not uniquely determine a matroid. However, we can construct a matroid from a given lattice.

5.2.1 DEFINITION: LATTICE OF FLATS

Let $M(S,T)$ be a matroid and let F be the set of flats of M. Let $\mathfrak{F}$ be a poset on F where the elements of F are ordered by inclusion. The poset $\mathfrak{F}$ forms a lattice called the lattice of flats of M.

The poset $\mathfrak{F}$ must form a lattice because M has a unique minimal and maximal element (so long as M is not the trivial matroid). The minimal element is $\emptyset$, and the maximal element is S. All elements of $\mathfrak{F}$ can be ordered with respect to these two unique elements.

The element $\emptyset$ is a member of F because if $X = cl(\emptyset)$, then X is the unique closure of elements or rank 0. This is the minimal element in $\mathfrak{F}$ because $\emptyset \subseteq A$ for every set A.

The element S is a member of F because $cl(S) = S$ for every matroid. This is the maximal element in $\mathfrak{F}$ because all elements of $\mathfrak{F}$ must be subsets of S.

The lattice arising from a matroid is sometimes referred to as a geometric lattice.

5.2.2 THEOREM

Let $M(S,T)$ be a matroid and I be a set that is not a flat. Then no superset of I is a flat.

Proof. If I is not a flat then $r(I) \neq |I|$. But it is never possible for $r(I) > |I|$, so we must have $r(I) < |I|$.

Let $e \in S$ be an element of S that is not already in I ($e \notin I$). We know

$$r(I \cup e) + r(I \cap e) \le r(I) + r(e).$$

But $I \cap e = \emptyset$ because e is not in I. Furthermore, $r(e) \le 1$. Substituting this into the previous inequality,

$$r(I \cup e) \le r(I) + 1.$$

Now, we know $|I \cup e| = |I| + 1$ because $e \notin I$. Moreover, since $r(I) < |I|$, we have

$$r(I \cup e) < |I| + 1 = |I \cup e|.$$

Therefore, $I \cup e$ is also not a flat.

We see that if we start with a set I that is not a flat, then we cannot create a flat by adding a single element to I. But since all of these new sets are also not flats, we cannot create a flat by adding another element to them either. By induction, no superset of I can be a flat. ■

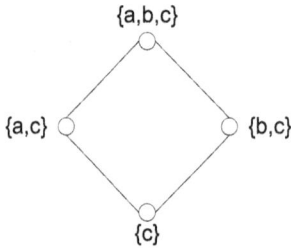

Figure 22: Lattice of flats for the matroid in example 5.2.3.

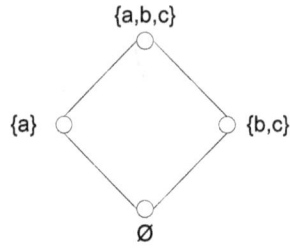

Figure 23: Lattice of flats for the matroid in example 5.2.4.

5.2.3 EXAMPLE

Let $M(S, B)$ be a matroid with ground set and basis

$$S = \{a, b, c\}$$
$$B = \{\{a, b\}\}.$$

Find the lattice of flats of M.

First, we need to find all flats of M. We systematically check every subset of S unless we find a subset that is not a flat. If we find a set that is not a flat, we do not need to check any of its supersets.

Rank 0 independent sets:

$$\emptyset \quad \backslash \quad \{c\}$$

Rank 1 independent sets:

$$\{a\} \quad \{b\} \quad \{a, c\} \quad \{b, c\}$$

Rank 2 independent sets:

$$\{a, b\} \quad \{a, b, c\}$$

For each rank group, we identify the maximal sets. These are the closure sets for each rank:

$$0 \quad \{c\}$$
$$1 \quad \{a,c\}\{b,c\}$$
$$2 \quad \{a,b,c\}$$

These are the flats of M. These sets can be ranked by inclusion because

$$\{c\} \subset \{a,c\} \subset \{a,b,c\}$$
$$\{c\} \subset \{b,c\} \subset \{a,b,c\}$$

This poset forms the lattice shown in Figure 22.

5.2.4   EXAMPLE

Let $M(S,B)$ be a matroid with ground set and basis

$$S = \{a,b,c\}$$
$$B = \{\{a,b\},\{a,c\}\}$$

Find the lattice of flats of M.

We proceed as in the previous example.

Rank 0 sets:

$$\emptyset$$

Rank 1 sets:

$$\{a\} \quad \{b\} \quad \{c\} \quad \{b,c\}$$

Rank 2 sets:

$$\{a,b\} \quad \{a,c\} \quad \{a,b,c\}$$

For each rank group, we identify the maximal sets. These are the closures groups for each rank:

$$0 \quad \emptyset$$
$$1 \quad \{a\}\{b,c\}$$
$$2 \quad \{a,b,c\}$$

These are the flats of M. These sets can be ranked by inclusion because

$$\emptyset \subset \{a\} \subset \{a, b, c\}$$
$$\emptyset \subset \{b, c\} \subset \{a, b, c\}`$$

This poset forms the lattice shown in Figure 23.

The lattices in the last examples are isomorphic to each other even though the underlying matroids are different. This is an example showing that a lattice does not uniquely determine a matroid. In this case, we see there are at least two different matroids that may arise from this lattice.

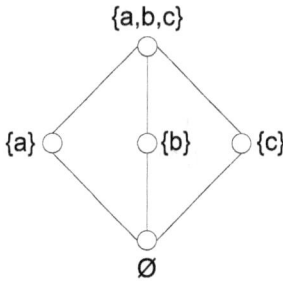

Figure 24: Lattice of flats for the matroid in example 5.2.5.

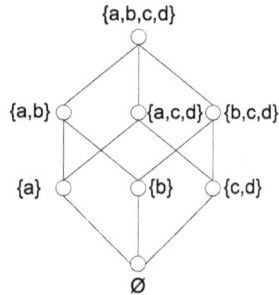

Figure 25: Lattice of flats for the matroid in example 5.2.6.

5.2.5   EXAMPLE

Let $M(S, B)$ be a matroid with ground set and basis

$$S = \{a, b, c\}$$
$$B = \{\{a, b\}, \{a, b\}, \{b, c\}\}$$

Find the lattice of flats of M.

We proceed as in the previous examples.

Rank 0 sets:

$$\emptyset$$

Rank 1 sets:

$$\{a\} \quad \{b\} \quad \{c\}$$

Rank 2 sets:

$$\{a,b\}\quad\{a,c\}\quad\{b,c\}\quad\{a,b,c\}$$

For each rank group, we identify the maximal sets. These are the closures groups for each rank:

$$
\begin{array}{ll}
0 & \emptyset \\
1 & \{a\}\{b\}\{c\} \\
2 & \{a,b,c\}
\end{array}
$$

These are the flats of M. These sets can be ranked by inclusion because

$$
\begin{array}{l}
\emptyset \subset \{a\} \subset \{a,b,c\} \\
\emptyset \subset \{b\} \subset \{a,b,c\} \\
\emptyset \subset \{c\} \subset \{a,b,c\}
\end{array}
$$

This poset forms the lattice shown in Figure 24.

5.2.6   EXAMPLE

Let $M(S,B)$ be a matroid with ground set and basis

$$
\begin{array}{l}
S = \{a,b,c,d\} \\
B = \{\{a,b,c\},\{a,b,d\}\}
\end{array}
$$

Find the lattice of flats of M.

Rank 0 sets:

$$\emptyset$$

Rank 1 sets:

$$\{a\}\quad\{b\}\quad\{c\}\quad\{d\}\quad\{c,d\}$$

Rank 2 sets:

$$\{a,b\}\quad\{a,c\}\quad\{a,d\}\quad\{b,c\}\quad\{b,d\}\quad\{a,c,d\}\quad\{b,c,d\}$$

Rank 3 sets:

$$\{a,b,c\}\{a,b,d\}\{a,b,c,d\}$$

For each rank group, we identify the maximal sets. These are the closures groups for each rank:

$$
\begin{array}{cl}
0 & \emptyset \\
1 & \{a\}\{b\}\{c,d\} \\
2 & \{a,b\}\{a,c,d\}\{b,c,d\} \\
3 & \{a,b,c,d\}
\end{array}
$$

The flats of M are diagrammed as a lattice in Figure 25.

5.3   Rank Generating Function

The rank generating function of a matroid is a two-variable polynomial created from the subsets of the ground set of the matroid. The rank generating function connects matroid theory with the powerful combinatorial machinery of generating functions. The rank generating function is an invariant and produces the same polynomial from every representation of a matroid.

5.3.1   DEFINITION: RANK GENERATING FUNCTION

Let M be a matroid with rank function r, ground set S, and lattice of flats $\mathfrak{L}$. The rank generating function of M is a polynomial in θ and ϕ defined as

$$
R_M(\theta, \phi) = \sum_{X \subseteq S} \theta^{r(M)-r(X)} \phi^{|X|-r(X)}
$$

The sum is over all subsets X of S, and $|X|$ is the cardinality of the subset X.

5.3.2   EXAMPLE

Compute the rank generating function for the matroid with ground set and basis given by

$$
S = \{a,b,c\}
$$
$$
B = \{\{a,b\},\{a,b\},\{b,c\}\}
$$

This is the matroid of example 5.2.5. The lattice of flats is

$$
\begin{array}{cl}
0 & \emptyset \\
1 & \{a\}\{b\}\{c\} \\
2 & \{a,b,c\}
\end{array}
$$

$$R_M(\theta, \phi) = \sum_{X \subseteq S} \theta^{r(M)-r(X)} \phi^{|X|-r(X)}$$

$$= \theta^{2-r(\emptyset)} \phi^{|\emptyset|-r(\emptyset)} + \theta^{2-r(\{a\})} \phi^{|\{a\}|-r(\{a\})} + \theta^{2-r(\{b\})} \phi^{|\{b\}|-r(\{b\})}$$
$$+ \theta^{2-r(\{c\})} \phi^{|\{c\}|-r(\{c\})} + \theta^{2-r(\{a,b\})} \phi^{|\{a,b\}|-r(\{a,b\})}$$
$$+ \theta^{2-r(\{a,c\})} \phi^{|\{a,c\}|-r(\{a,c\})} + \theta^{2-r(\{b,c\})} \phi^{|\{b,c\}|-r(\{b,c\})}$$
$$+ \theta^{2-r(\{a,b,c\})} \phi^{|\{a,b,c\}|-r(\{a,b,c\})}$$

$$= \theta^{2-0} \phi^{0+0} + \theta^{2-1} \phi^{1-1} + \theta^{2-1} \phi^{1-1} + \theta^{2-1} \phi^{1-1} + \theta^{2-2} \phi^{2-2}$$
$$+ \theta^{2-2} \phi^{2-2} + \theta^{2-2} \phi^{2-2} + \theta^{2-2} \phi^{3-2}$$

$$= \theta^2 + 3\theta + 3 + \phi$$

5.3.3   PROPOSITION

Let M be a matroid with rank function r, ground set S, set of bases B, independent set I, and lattice of flats $\mathfrak{L}$. The rank generating function of M has the properties

RG-1 $\mu_M(\emptyset, S) = (-1)^{r(M)} R_M(0, -1)$

RG-2 $\chi_M(\lambda) = (-1)^{r(M)} R_M(-\lambda, -1)$

RG-3 $|I| = R_M(1, 0)$

RG-4 $|B| = R_M(0, 0)$

Proof. The first result follows from **MI-1**. The second result follows from the definition of the rank polynomial and rank generating function. The third and fourth results follow from **MI-1**, with **RP-1** and **RP-2**.

From **MI-1**,

$$\mu_M(A, B) = \sum_{\substack{A \subseteq X \subseteq B \\ cl(X)=B}} (-1)^{|X-A|}$$

Setting $A = \emptyset$ and $B = S$, and recognizing that the resulting sum is over all subsets of S,

$$\mu_M(\emptyset, S) = \sum_{X \subseteq S} (-1)^{|X|}$$

If we set $\theta = 0$ in the rank generating function, all terms will evaluate to zero except those where $r(M) = r(X)$. So,

$$R_M(0, \phi) = \sum_{X \subseteq S} \phi^{|X|-r(M)} = \phi^{r(M)} \sum_{X \subseteq S} \phi^{|X|}$$

If we also set $\phi = -1$,

$$R_M(0, -1) = (-1)^{r(M)} \sum_{X \subseteq S} (-1)^{|X|}$$

With this we can write the Möbius invariant in terms of the rank generating function as

$$\mu_M(\emptyset, S) = (-1)^{r(M)} R_M(0, -1)$$

For the second result, we see

$$R_M(-\lambda, -1) = \sum_{X \subseteq S} (-\lambda)^{r(M)-r(X)} (-1)^{|X|-r(X)}$$

$$= \sum_{X \subseteq S} (-1)^{r(M)-r(X)} (-1)^{|X|-r(X)} \lambda^{r(M)-r(X)}$$

$$= \sum_{X \subseteq S} (-1)^{r(M)+|X|-2r(X)} \lambda^{r(M)-r(X)}$$

$$= \sum_{X \subseteq S} (-1)^{r(M)+|X|} \lambda^{r(M)-r(X)}$$

$$= (-1)^{r(M)} \sum_{X \subseteq S} (-1)^{|X|} \lambda^{r(M)-r(X)}$$

$$= (-1)^{r(M)} \chi_M(\lambda)$$

where the last step is from **CP-1**.

For the third result, all terms of $R_M(1,0)$ will be zero unless $|X| = r(X)$. From **RP-1**, if $|X| = r(X)$, then X is an independent set. In this case, we get one term for each independent set in M.

Moreover, because $\theta = 1$, each of these terms evaluates to 1. Thus, we have one 1 for every independent set. So,

$$|I| = R_M(1,0)$$

The fourth result is similar to the third. In the case of $R_M(0,0)$, all terms will be zero unless $|X| = r(X) = r(M)$. From RP-2, this is true only if X is a basis of M. Therefore, $R_M(0,0)$ counts the number of basis elements in M:

$$|B| = R_M(0,0) \blacksquare$$

5.4   Rank Polynomial

The rank polynomial of a matroid is a simple polynomial created from the flats of the matroid. Each term of the polynomial arises from one of the flats, and the exponent on the variable equals to the rank of the matroid minus the rank of the flat.

5.4.1   DEFINITION: RANK POLYNOMIAL

Let M be a matroid with rank function r and lattice of flats $\mathfrak{L}$. The rank polynomial of M is a polynomial in λ defined as

$$\rho_M(\lambda) = \sum_{F \in \mathfrak{L}} \lambda^{r(M)-r(F)}$$

5.4.2   EXAMPLE

Compute the rank polynomial for the matroid with ground set and basis given by

$$S = \{a,b,c\}$$
$$B = \{\{a,b\},\{a,b\},\{b,c\}\}'$$

This is the matroid of example 5.2.5. The lattice of flats is

$$
\begin{array}{ll}
0 & \emptyset \\
1 & \{a\}\{b\}\{c\} \\
2 & \{a,b,c\}
\end{array}
$$

$$\rho_M(\lambda) = \sum_{F \in \mathfrak{L}} \lambda^{r(M)-r(F)} = \lambda^{2-0} + \lambda^{2-1} + \lambda^{2-1} + \lambda^{2-1} + \lambda^{2-2}$$

$$= \lambda^2 + 3\lambda + 1$$

5.4.3   EXAMPLE

Compute the characteristic polynomial for the matroid with ground set and basis given by

$$S = \{a, b, c, d\}$$
$$B = \{\{a, b, c\}, \{a, b, d\}\}$$

This is the matroid of example 5.2.6. The lattice of flats is

0	$\emptyset$
1	$\{a\}\{b\}\{c, d\}$
2	$\{a, b\}\{a, c, d\}\{b, c, d\}$
3	$\{a, b, c, d\}$

$$\rho_M(\lambda) = \sum_{F \in \mathfrak{L}} \lambda^{r(M)-r(F)}$$

$$= \lambda^{3-0} + \lambda^{3-1} + \lambda^{3-1} + \lambda^{3-1} + \lambda^{3-2} + \lambda^{3-2} + \lambda^{3-2} + \lambda^{3-3}$$

$$= \lambda^3 + 3\lambda^2 + 3\lambda + 1$$

5.4.4   PROPOSITION

Let M be a matroid with rank polynomial $\rho_M(\lambda)$. The order of $\rho_M(\lambda)$ is the rank of M.

Proof. First, we show that there is always a term of order $r(M)$. From the definition of $\rho_M(\lambda)$, we have terms of order $r(M)$ if there is a flat with $r(F) = 0$. Furthermore, because all terms have positive coefficients, so long as we have one term, the final polynomial must have a term of the same order with a non-zero coefficient.

For any matroid, there is always a flat with rank $r(F) = 0$. By definition $r(\emptyset) = 0$. If $\emptyset$ is a flat, this is satisfied. If $\emptyset$ is not a flat, we can find a flat from the closure of $\emptyset$. This flat must have rank 0 by definition of closure. We see there is always at least one flat with $r(F) = 0$.

Second, we show there is no terms with order higher than $r(M)$. Assume the contrary. Then there must be a term of order x such that $x > r(M)$.

From the definition of the rank polynomial, this term came from a flat with the property

$$x = r(M) - r(f)$$

or

$$r(f) = r(M) - x$$

But since $x > r(M)$, we have $r(f) < 0$. This is contrary to the definition of rank. Thus, there are no terms with order higher than $r(M)$.∎

5.4.5   RANK POLYNOMIAL PROPERTIES

The following properties follow immediately from the properties of matroids and definition of the rank polynomial:

RP-1 $\rho_M(0) = 1$

RP-2 $\rho_M(1) = |\mathfrak{L}|$

5.5   Zeta and Möbius Functions

The zeta function is defined over a poset and is related to the cardinality of the poset. In fact, the zeta function measures set-subset relationships. This counting application is a useful measure in many combinatorial applications.

The Möbius function is the inverse of the zeta function. The Möbius function provides a foundation to define the characteristic polynomial. We will later see that the characteristic polynomial for matroids, based on the Möbius function, is similar to the chromatic polynomial for graphs.

5.5.1 DEFINITION: ZETA FUNCTION

Let P be a finite poset. The zeta function is defined as

$$\zeta(x, y) = \begin{cases} 1 & x \leq y \\ 0 & otherwise \end{cases}.$$

It is important to note that the value for the zeta function is 1 when $x \leq y$ and 0 otherwise. This is not equivalent to 0 when $x > y$ because the elements of a poset may not be comparable. If two element are not comparable, their zeta function evaluates to 0.

5.5.2 EXAMPLE

Let $\mathfrak{L}$ be the lattice of Figure 26. Find the zeta function for this lattice.

The zeta function is a relation between pairs of values. We specify the values of the zeta function as a table where the columns are the x-values and the rows are the y-values:

	a	b	c	d	e
a	1	1	1	1	1
b	0	1	0	1	1
c	0	0	1	1	1
d	0	0	0	1	1
e	0	0	0	0	1

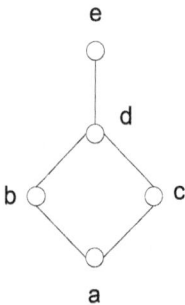

Figure 26: Sample lattice with five elements.

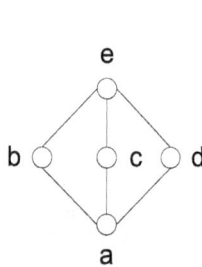

Figure 27: Another lattice with five elements.

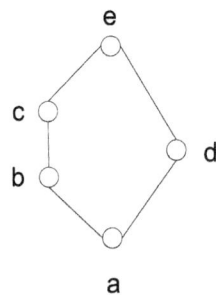

Figure 28: Third lattice with five elements.

It should be noted that $\zeta(b,c) = \zeta(c,b) = 0$. Looking at the diagram in Figure 26, the elements b and c are not comparable, hence their zeta function evaluates to 0.

5.5.3   EXAMPLE

Let $\mathcal{L}$ be the lattice of Figure 27. Find the zeta function for this lattice.

Similar to the previous example, we specify the values of the zeta function as a table where the columns are the x-values and the rows are the y-values:

	a	b	c	d	e
a	1	1	1	1	1
b	0	1	0	0	1
c	0	0	1	0	1
d	0	0	0	1	1
e	0	0	0	0	1

5.5.4   EXAMPLE

Let $\mathcal{L}$ be the lattice of Figure 28. Find the zeta function for this lattice.

Similar to the previous examples, we specify the values of the zeta function as a table where the columns are the x-values and the rows are the y-values:

	a	b	c	d	e
a	1	1	1	1	1
b	0	1	1	0	1
c	0	0	1	0	1
d	0	0	0	1	1
e	0	0	0	0	1

5.5.5 DEFINITION: INCIDENCE ALGEBRA

> Let P be a finite poset. Let $f(x,y)$ and $g(x,y)$ be functions
> mapping $P \times P \to \mathbb{Z}$. The convolution of functions in the algebra is
> defined as
>
> $$f \circ g = fg(x,y) = \sum_{x \leq z \leq y} f(x,z)g(z,y).$$
>
> The algebra of functions over P is called the incidence algebra.

5.5.6 EXAMPLE

Compute $\zeta^2(x,y)$ for the lattice in Figure 26.

$$\zeta^2(x,y) = \sum_{x \leq z \leq y} \zeta(x,z)\zeta(z,y).$$

Let's look at a few specific cases.

$$\zeta^2(x,x) = \sum_{x \leq z \leq x} \zeta(x,z)\zeta(z,x) = \zeta(x,x)\zeta(x,x) = 1.$$

There is only one term in this sum, the term with $z = x$. Since
$\zeta(x,x) = 1$ for every value of x, we have $\zeta^2(x,x) = 1$.

$$\zeta^2(a,b) = \sum_{a \leq z \leq b} \zeta(a,z)\zeta(z,b) = \zeta(a,a)\zeta(a,b) + \zeta(a,b)\zeta(b,b) = 2.$$

$$\zeta^2(a,d) = \sum_{a \leq z \leq d} \zeta(a,z)\zeta(z,d)$$
$$= \zeta(a,a)\zeta(a,d) + \zeta(a,b)\zeta(b,d) + \zeta(a,c)\zeta(c,d)$$
$$+ \zeta(a,d)\zeta(d,d) = 4.$$

$$\zeta^2(a,e) = \sum_{a \leq z \leq e} \zeta(a,z)\zeta(z,e)$$
$$= \zeta(a,a)\zeta(a,e) + \zeta(a,b)\zeta(b,e) + \zeta(a,c)\zeta(c,e)$$
$$+ \zeta(a,d)\zeta(d,e) + \zeta(a,e)\zeta(e,e) = 5.$$

We express the entire results of the function as a table:

	a	b	c	d	e
a	1	2	2	4	5
b	0	1	0	2	3
c	0	0	1	3	4
d	0	0	0	1	2
e	0	0	0	0	1

It is interesting to note that the value of $\zeta^2(x, y)$ is the cardinality ($card$) of the members of the poset contained on the range $[x, y]$.

5.5.7   PROPOSITION

Let P be a poset and $\zeta(x, y)$ the zeta function on the poset. Then

$$\zeta^2(x, y) = card([x, y]).$$

Proof. From the definition of the zeta function we know

$$\zeta^2(x, y) = \sum_{x \leq z \leq y} \zeta(x, z)\zeta(z, y).$$

In the case where $x \geq y$, there is no z such that $x \leq z \leq y$. Because no z satisfies this inequality, there are no terms in the sum so $\zeta^2(x, y) = 0$.

In the case where x and y are not comparable, again, there is no z such that $x \leq z \leq y$. Similarly, $\zeta^2(x, y) = 0$.

In the case where $x \leq y$, suppose there are k values of z such that $x \leq z \leq y$. Each of these will contribute a term to the sum in the form $\zeta(x, z_k)\zeta(z_k, y)$. But since $x \leq z_k$, $\zeta(x, z_k) = 1$. Moreover, since $z_k \leq y$, $\zeta(z_k, y) = 1$.

We see that each of the k values contributes 1 to the sum. Thus,

$$\zeta^2(x, y) = k.$$

Putting these results together we see that in each case $\zeta^2(x,y)$ matches the cardinality of the poset elements satisfying the order relation. Thus,

$$\zeta^2(x,y) = card([x,y]).$$

5.5.8   PROPOSITION

Let P be a poset and $\zeta(x,y)$ the zeta function on the poset. Then the number of chains of length k between x and y is

$$(\zeta - 1)^k(x,y).$$

Proof (Sketch). Start by examining $(\zeta - 1)(x,y)$. This equation is written in a functional sense, meaning

$$(\zeta - 1)(x,y) \equiv \zeta(x,y) - \delta(x,y).$$

This function evaluates to

$$(\zeta - 1)(x,y) = \zeta(x,y) - \delta(x,y) = \begin{cases} 1 & x < y \\ 0 & otherwise \end{cases}.$$

This function will correctly provide the number of 1-dimensional chains because there can only be one such chain, and the chain is only present if $x < y$ (remember, when counting the number of chains of length n, we count the number of possible ways we can formulate $x_0 < x_1 < x_2 < \cdots < x_n$).

Furthermore,

$$(\zeta - 1)^2(x,y) \equiv \zeta^2(x,y) - 2\zeta(x,y) + \delta(x,y)$$

$$= \begin{cases} card[x,y] - 2 & x < y \\ 0 & otherwise \end{cases}.$$

The total number of chains of length 2 is the total number of elements satisfying $x < z < y$. This is simply the number of elements in the poset between x and y. Remember, the cardinality of $[x,y]$ is the total number of such elements *including* x and y. Thus, the total number of chains of length 2 is $card[x,y] - 2$ if $x < y$, and 0 otherwise.

Similar results hold for all k yielding $(\zeta - 1)^k(x,y)$ as the number of chains of length k. ∎

5.5.9 DEFINITION: MÖBIUS FUNCTION

Let P be a finite poset. Let $\mu(x,y)$ be the inverse of the zeta function in the sense

$$\mu \circ \zeta = \mu\zeta(x,y) = \sum_{x \leq z \leq y} \mu(x,z)\zeta(z,y) = \delta(x,y).$$

Specifically,

$$\mu(x,y) = \begin{cases} 0 & x \not\leq y \\ 1 & x = y \\ -\sum_{x \leq z < y} \mu(x,z) & \forall\, x < y \in P \end{cases}$$

The function $\delta(x,y)$ is the Kronecker delta function

$$\delta(x,y) = \begin{cases} 1 & x = y \\ 0 & x \neq y \end{cases}.$$

The second definition is obtained from the first using recursion. For example, if $x \not\leq y$ then the zeta function is zero. Furthermore, if $x = y$

$$\mu \circ \zeta(x,x) = \sum_{x \leq z \leq y} \mu(x,x)\zeta(x,x) = \mu(x,x)\zeta(x,x) = \mu(x,x) = 1.$$

Next, assume we have $x < y$ where there are no elements $x < z < y$. In this case

$$\mu \circ \zeta(x,y) = \sum_{x \leq z \leq y} \mu(x,z)\zeta(z,x) = \mu(x,x)\zeta(x,y) + \mu(x,y)\zeta(y,y)$$
$$= 1 + \mu(x,y) = 0$$

So,

$$\mu(x,y) = -1 = -\mu(x,x) = -\sum_{x \leq z < y} \mu(x,z).$$

Now assume there is exactly one z such that $x < z < y$. We have

$$\mu \circ \zeta(x,y) = \sum_{x \le z \le y} \mu(x,z)\zeta(z,x)$$
$$= \mu(x,x)\zeta(x,y) + \mu(x,z)\zeta(z,y) + \mu(x,y)\zeta(y,y)$$
$$= 1 + \mu(x,z) + (-1) = 0$$

or,

$$\mu(x,y) = 0 = -1 - (-1) = -\sum_{x \le z < y}\mu(x,z).$$

If we continue this process we arrive at the second definition stated above.

5.5.10 EXAMPLE

Compute $\mu(x,y)$ for the lattice in Figure 27.

First, we know $\mu(x,x) = 1$ for every element. Next, compute $\mu(a,b)$:

$$\mu(a,b) = -\sum_{a \le z < b}\mu(a,z) = -\mu(a,a) = -1.$$

Similarly,

$$\mu(a,c) = -\sum_{a \le z < c}\mu(a,z) = -\mu(a,a) = -1.$$

$$\mu(a,d) = -\sum_{a \le z < b}\mu(a,z) = -\mu(a,a) = -1.$$

Moreover,

$$\mu(a,e) = -\sum_{x \le z < y}\mu(a,z) = -\mu(a,a) - \mu(a,b) - \mu(a,c) - \mu(a,d)$$
$$= -1 - (-1) - (-1) - (-1) = 2.$$

Finally,

$$\mu(b,e) = -\sum_{b \le z < e}\mu(a,z) = -\mu(b,b) = -1.$$

$$\mu(c,e) = -\sum_{c \le z < e}\mu(a,z) = -\mu(b,b) = -1.$$

$$\mu(d,e) = - \sum_{d \leq z < e} \mu(a,z) = -\mu(b,b) = -1.$$

Putting these results into a table,

	a	b	c	d	e
a	1	−1	−1	−1	2
b	0	1	0	0	−1
c	0	0	1	0	−1
d	0	0	0	1	−1
e	0	0	0	0	1

5.5.11 EXAMPLE

Compute $\mu(x,y)$ for the lattice in Figure 28.

We know $\mu(x,x) = 1$. Furthermore, $\mu(x,y) = 0$ unless $x \leq y$. Because of this we only need to compute the off diagonal elements with $x < y$.

$$\mu(a,b) = - \sum_{a \leq z < b} \mu(a,z) = -\mu(a,a) = -1.$$

$$\mu(a,c) = - \sum_{a \leq z < c} \mu(a,z) = -\mu(a,a) - \mu(a,b) = -1 - (-1) = 0.$$

$$\mu(a,d) = - \sum_{a \leq z < d} \mu(a,z) = -\mu(a,a) = -1.$$

$$\mu(a,e) = - \sum_{a \leq z < e} \mu(a,z) = -\mu(a,a) - \mu(a,b) - \mu(a,c) - \mu(a,d)$$
$$= -1 - (-1) - 0 - (-1) = 1.$$

$$\mu(b,c) = - \sum_{b \leq z < c} \mu(b,z) = -\mu(b,b) = -1.$$

$$\mu(b,e) = - \sum_{b \leq z < e} \mu(b,z) = -\mu(b,b) - \mu(b,c) = -1 - (-1) = 0$$

$$\mu(c,e) = -\sum_{c \leq z < e} \mu(c,z) = -\mu(c,c) = -1.$$

$$\mu(d,e) = -\sum_{d \leq z < e} \mu(d,z) = -\mu(d,d) = -1.$$

In the form of a table,

	a	b	c	d	e
a	1	−1	0	−1	1
b	0	1	−1	0	0
c	0	0	1	0	−1
d	0	0	0	1	−1
e	0	0	0	0	1

Again, we see that the Möbius function is 0 whenever $x > y$ or when x and y are not comparable. However, the Möbius function may be zero when $x \leq y$. For example, $\mu(b,e) = 0$ even though $b < e$.

5.5.12 DEFINITION: MÖBIUS INVERSION

Let P be a finite poset. Let $f(x)$ and $g(x)$ be functions on P mapping P to a ring ($\mathbb{Z}, \mathbb{Q}, \mathbb{R}, \mathbb{C}, \mathbb{H}$, etc.), where $f(x)$ and $g(x)$ satisfy

$$g(x) = \sum_{y \geq x} f(y)$$

then

$$f(x) = \sum_{y \geq x} \mu(x,y)g(y).$$

Similarly, it is true that if

$$g(x) = \sum_{y \leq x} f(y)$$

then

$$f(x) = \sum_{y \leq x} \mu(x,y)g(y).$$

The Möbius inversion formula provides a means to invert a specific relation between arbitrary functions on P. The inversion formula follows directly from the definition of the Möbius function. Specifically, since $\mu \circ \zeta = \delta$, we have $f = (\mu \circ \zeta) \circ f$.

However, the incidence algebra is associative, so $(\mu \circ \zeta) \circ f = \mu \circ (\zeta \circ f)$. Thus, $f = \mu \circ (\zeta \circ f) = \mu \circ g$, where $g = \zeta \circ f$.

Alternatively, if we start with $\mu \circ \zeta = \delta$, we can formally invert this expression to obtain

$$\mu = \zeta^{-1}(x, y) = \left(1 + (\zeta - 1)\right)^{-1}(x, y)$$

$$= \delta(x, y) - (\zeta - 1)(x, y) + (\zeta - 1)^2(x, y) - \cdots$$

$$= c_0 - c_1 + c_2 - \cdots$$

where c_n is the number of chains of length n between x and y.

5.5.13 DEFINITION: EULER CHARACTERISTIC

> Let P be a finite poset with a minimum element $\hat{0}$ and a maximum element $\hat{1}$. The Euler characteristic of P is the value of the Möbius function
>
> $$\chi = \mu_P(\hat{0}, \hat{1})$$
>
> Where the subscript on μ indicates the poset under consideration.

5.5.14 EXAMPLE

Compute the Euler characteristic for the lattices in Figure 26-Figure 28.

In each case, the Euler characteristic is $\chi = \mu_P(a, e)$. For Figure 26,

$$\mu_1(a, e) = -\sum_{a \leq z < e} \mu(a, z)$$
$$= -\mu(a, a) - \mu(a, b) - \mu(a, c) - \mu(a, d).$$

We know $\mu(a, a)$. We need to compute the remaining values:

$$\mu_1(a,b) = - \sum_{a \leq z < b} \mu(a,z) = -\mu(a,a) = -1.$$

$$\mu_1(a,c) = - \sum_{a \leq z < c} \mu(a,z) = -\mu(a,a) = -1.$$

$$\mu_1(a,d) = - \sum_{a \leq z < d} \mu(a,z) = -\mu(a,a) - \mu(a,b) - \mu(a,c)$$
$$= -1 - (-1) - (-1) = 1.$$

Putting everything together,

$$\chi_1 = \mu_1(a,e) = -1 - (-1) - (-1) - 1 = 0$$

The other two are computed similarly. However, we have already computed the value of the Möbius function for these lattices in previous examples. Reading the appropriate values from:

$$\chi_2 = 2 \quad \chi_3 = 1$$

5.6   Möbius Invariant

The Möbius function may be extended to matroids using the matroids lattice of flats. The lattice of flats of a matroid provide a poset that can be used in the Möbius function.

Applying the machinery of the Möbius function to matroids, we can extend the concept of the Euler characteristic. This is the matroid Möbius invariant, which we will later use to extend the concept of the chromatic polynomial of a graph.

5.6.1   DEFINITION: MÖBIUS FUNCTION OF A MATROID

Let M be a matroid with lattice of flats $\mathfrak{L}$. The Moëbius function of M is based on the Moëbius function for $\mathfrak{L}$:

$$\mu_M(X,F) = \begin{cases} \mu_{\mathfrak{L}}(X,F) & X,F \in \mathfrak{L} \\ 0 & othervise \end{cases}$$

5.6.2   PROPOSITION

Let M be a matroid with lattice of flats $\mathfrak{L}$. The Möbius function $\mu_M(X,F)$ is 0 if $X \nsubseteq F$.

Proof. From the definition of the Möbius function, $\mu_M(X, F) = 0$ unless $X, F \in \mathfrak{L}$. In this case, the Möbius function is

$$\mu_{\mathfrak{L}}(X, F) = \begin{cases} 0 & X \not\leq F \\ 1 & X = F \\ -\sum_{X \leq z < F} \mu_{\mathfrak{L}}(X, z) & \forall \, X < F \in \mathfrak{L} \end{cases}$$

The lattice $\mathfrak{L}$ is ordered by inclusion. This means that $z < F$ if $z \subset F$, and $X \leq z$ if $X \subseteq z$.

It is useful to rewrite the Möbius function in terms of the subset operator as

$$\mu_{\mathfrak{L}}(X, F) = \begin{cases} 0 & X \not\subseteq F \\ 1 & X = F \\ -\sum_{X \subseteq z \subset F} \mu_{\mathfrak{L}}(X, z) & \forall \, X \subset F \in \mathfrak{L} \end{cases}$$

The first statement is the desired result. ∎

5.6.3   EXAMPLE

Let $M(S, B)$ be a matroid with ground set and basis

$$S = \{a, b, c\}$$
$$B = \{\{a, b\}\}.$$

Compute the Möbius function for this matroid.

This is the matroid from section 5.2.3 with lattice of flats $\mathfrak{L}$

$$\begin{array}{ll} 0 & \{c\} \\ 1 & \{a, c\}\{b, c\} \\ 2 & \{a, b, c\} \end{array}$$

The Möbius function is zero unless both $X, F \in \mathfrak{L}$. With four elements, there are sixteen choices for combinations of X, F. We examine each case separately:

$X = \{c\} \quad F = \{c\}$

In this case $X = F$, so $\mu_\varrho(X, F) = 1$.

$X = \{c\} \quad F = \{a, c\}$

$$\mu_\varrho(X, F) = -\sum_{X \subseteq z \subset F} \mu_\varrho(X, z) = -\mu_\varrho(\{c\}, \{c\}) = -1$$

$X = \{c\} \quad F = \{b, c\}$

$$\mu_\varrho(X, F) = -\sum_{X \subseteq z \subset F} \mu_\varrho(X, z) = -\mu_\varrho(\{c\}, \{c\}) = -1$$

$X = \{c\} \quad F = \{a, b, c\}$

$$\mu_\varrho(X, F) = -\sum_{X \subseteq z \subset F} \mu_\varrho(X, z)$$
$$= -\mu_\varrho(\{c\}, \{c\}) - \mu_\varrho(\{c\}, \{a, c\}) - \mu_\varrho(\{c\}, \{b, c\})$$
$$= -1 - (-1) - (-1) = 1$$

$X = \{a, c\} \quad F = \{c\}$

Here, $X \not\subseteq F$, so $\mu_\varrho(X, F) = 0$.

$X = \{a, c\} \quad F = \{a, c\}$

$$\mu_\varrho(X, F) = 1$$

$X = \{a, c\} \quad F = \{b, c\}$

Again, $X \not\subseteq F$, so $\mu_\varrho(X, F) = 0$.

$X = \{a, c\} \quad F = \{a, b, c\}$

$$\mu_\varrho(X, F) = -\sum_{X \subseteq z \subset F} \mu_\varrho(X, z) = -\mu_\varrho(\{a, c\}, \{a, c\}) = -1$$

$X = \{b, c\} \quad F = \{c\}$

Here, $X \not\subseteq F$, so $\mu_\varrho(X, F) = 0$.

$X = \{b, c\}$ $F = \{b, c\}$

Again, $X \nsubseteq F$, so $\mu_{\varrho}(X, F) = 0$.

$X = \{b, c\}$ $F = \{b, c\}$

$$\mu_{\varrho}(X, F) = 1$$

$X = \{b, c\}$ $F = \{a, b, c\}$

$$\mu_{\varrho}(X, F) = - \sum_{X \subseteq z \subset F} \mu_{\varrho}(X, z) = -\mu_{\varrho}(\{b, c\}, \{b, c\}) = -1$$

$X = \{a, b, c\}$ $F = \{c\}$

Here, $X \nsubseteq F$, so $\mu_{\varrho}(X, F) = 0$.

$X = \{a, b, c\}$ $F = \{b, c\}$

Again, $X \nsubseteq F$, so $\mu_{\varrho}(X, F) = 0$.

$X = \{a, b, c\}$ $F = \{b, c\}$

Again, $X \nsubseteq F$, so $\mu_{\varrho}(X, F) = 0$.

$X = \{a, b, c\}$ $F = \{a, b, c\}$

$$\mu_{\varrho}(X, F) = 1$$

We present the complete results as a table:

$X \backslash F$	$\{c\}$	$\{a, c\}$	$\{b, c\}$	$\{a, b, c\}$
$\{c\}$	1	−1	−1	1
$\{a, c\}$	0	1	0	−1
$\{b, c\}$	0	0	1	−1
$\{a, b, c\}$	0	0	0	1

5.6.4   EXAMPLE

Let $M(S, B)$ be a matroid with ground set and basis

$$S = \{a, b, c\}$$
$$B = \{\{a, b\}, \{a, c\}, \{b, c\}\}$$

Compute the Möbius function for this matroid.

This is the matroid from section 5.2.5 with lattice of flats $\mathfrak{L}$

$$
\begin{array}{cl}
0 & \emptyset \\
1 & \{a\}\{b\}\{c\} \\
2 & \{a, b, c\}
\end{array}
$$

In this case there are twenty-five combinations of X and F. Of these, there are only seven cases where $X \subset F$ (the cases where $X = F$ are trivial because $\mu_{\mathfrak{L}}(X, X) = 1$).

$X = \emptyset$ $F = \{a\}$

$$\mu_{\mathfrak{L}}(X, F) = - \sum_{X \subseteq z \subset F} \mu_{\mathfrak{L}}(X, z) = -\mu_{\mathfrak{L}}(\emptyset, \emptyset) = -1$$

$X = \emptyset$ $F = \{b\}$

$$\mu_{\mathfrak{L}}(X, F) = - \sum_{X \subseteq z \subset F} \mu_{\mathfrak{L}}(X, z) = -\mu_{\mathfrak{L}}(\emptyset, \emptyset) = -1$$

$X = \emptyset$ $F = \{c\}$

$$\mu_{\mathfrak{L}}(X, F) = - \sum_{X \subseteq z \subset F} \mu_{\mathfrak{L}}(X, z) = -\mu_{\mathfrak{L}}(\emptyset, \emptyset) = -1$$

$X = \emptyset$ $F = \{a, b, c\}$

$$\mu_{\mathfrak{L}}(X, F) = - \sum_{X \subseteq z \subset F} \mu_{\mathfrak{L}}(X, z)$$
$$= -\mu_{\mathfrak{L}}(\emptyset, \emptyset) - \mu_{\mathfrak{L}}(\emptyset, \{a\}) - \mu_{\mathfrak{L}}(\emptyset, \{b\})$$
$$- \mu_{\mathfrak{L}}(\emptyset, \{c\}) = -1 - (-1) - (-1) - (-1) = 2$$

$X = \{a\}$ $F = \{a, b, c\}$

$$\mu_{\mathfrak{L}}(X, F) = - \sum_{X \subseteq z \subset F} \mu_{\mathfrak{L}}(X, z) = -\mu_{\mathfrak{L}}(\{a\}, \{a\}) = -1$$

$X = \{b\}$ $F = \{a, b, c\}$

$$\mu_{\mathfrak{L}}(X, F) = - \sum_{X \subsetneqq z \subset F} \mu_{\mathfrak{L}}(X, z) = -\mu_{\mathfrak{L}}(\{b\}, \{b\}) = -1$$

$X = \{c\}$ $F = \{a, b, c\}$

$$\mu_{\mathfrak{L}}(X, F) = - \sum_{X \subsetneqq z \subset F} \mu_{\mathfrak{L}}(X, z) = -\mu_{\mathfrak{L}}(\{c\}, \{c\}) = -1$$

The table below presents the values for the Möbius function for the matroid:

$X \backslash F$	$\emptyset$	$\{a\}$	$\{b\}$	$\{c\}$	$\{a, b, c\}$
$\emptyset$	1	−1	−1	−1	2
$\{a\}$	0	1	0	0	−1
$\{b\}$	0	0	1	0	−1
$\{c\}$	0	0	0	1	−1
$\{a, b, c\}$	0	0	0	0	1

5.6.5 EXAMPLE

Let $M(S, B)$ be a matroid with ground set and basis

$$S = \{a, b, c, d\}$$
$$B = \{\{a, b, c\}, \{a, b, d\}\}$$

Compute the Möbius function for this matroid.

This is the matroid from section 5.2.6 with lattice of flats $\mathfrak{L}$

0	$\emptyset$
1	$\{a\}\{b\}\{c, d\}$
2	$\{a, b\}\{a, c, d\}\{b, c, d\}$
3	$\{a, b, c, d\}$

There are sixty four combinations of X and F, and nineteen cases where $X \subset F$.

$X = \emptyset \quad F = \{a\}$

$$\mu_\varrho(X, F) = -\sum_{X \subseteq z \subset F} \mu_\varrho(X, z) = -\mu_\varrho(\emptyset, \emptyset) = -1$$

$X = \emptyset \quad F = \{b\}$

$$\mu_\varrho(X, F) = -\sum_{X \subseteq z \subset F} \mu_\varrho(X, z) = -\mu_\varrho(\emptyset, \emptyset) = -1$$

$X = \emptyset \quad F = \{c, d\}$

$$\mu_\varrho(X, F) = -\sum_{X \subseteq z \subset F} \mu_\varrho(X, z) = -\mu_\varrho(\emptyset, \emptyset) = -1$$

$X = \emptyset \quad F = \{a, b\}$

$$\mu_\varrho(X, F) = -\sum_{X \subseteq z \subset F} \mu_\varrho(X, z)$$
$$= -\mu_\varrho(\emptyset, \emptyset) - \mu_\varrho(\emptyset, \{a\}) - \mu_\varrho(\emptyset, \{b\})$$
$$= -1 - (-1) - (-1) = 1$$

$X = \emptyset \quad F = \{a, c, d\}$

$$\mu_\varrho(X, F) = -\sum_{X \subseteq z \subset F} \mu_\varrho(X, z)$$
$$= -\mu_\varrho(\emptyset, \emptyset) - \mu_\varrho(\emptyset, \{a\}) - \mu_\varrho(\emptyset, \{c, d\})$$
$$= -1 - (-1) - (-1) = 1$$

$X = \emptyset \quad F = \{b, c, d\}$

$$\mu_\varrho(X, F) = -\sum_{X \subseteq z \subset F} \mu_\varrho(X, z)$$
$$= -\mu_\varrho(\emptyset, \emptyset) - \mu_\varrho(\emptyset, \{b\}) - \mu_\varrho(\emptyset, \{c, d\})$$
$$= -1 - (-1) - (-1) = 1$$

$X = \emptyset \quad F = \{a, b, c, d\}$

$$\mu_\varrho(X, F) = -\sum_{X \subseteq z \subset F} \mu_\varrho(X, z)$$
$$= -\mu_\varrho(\emptyset, \emptyset) - \mu_\varrho(\emptyset, \{a\}) - \mu_\varrho(\emptyset, \{b\})$$
$$- \mu_\varrho(\emptyset, \{c, d\}) - \mu_\varrho(\emptyset, \{a, c, d\}) - \mu_\varrho(\emptyset, \{b, c, d\})$$
$$= -1 - (-1) - (-1) - (-1) - 1 - 1 - 1 = -1$$

$X = \{a\}$ $F = \{a, b\}$

$$\mu_\varrho(X, F) = -\sum_{X \subseteq z \subset F} \mu_\varrho(X, z) = -\mu_\varrho(\{a\}, \{a\}) = -1$$

$X = \{a\}$ $F = \{a, c, d\}$

$$\mu_\varrho(X, F) = -\sum_{X \subseteq z \subset F} \mu_\varrho(X, z) = -\mu_\varrho(\{a\}, \{a\}) = -1$$

$X = \{a\}$ $F = \{a, b, c, d\}$

$$\mu_\varrho(X, F) = -\sum_{X \subseteq z \subset F} \mu_\varrho(X, z)$$
$$= -\mu_\varrho(\{a\}, \{a\}) - \mu_\varrho(\{a\}, \{a, b\})$$
$$- \mu_\varrho(\{a\}, \{a, c, d\}) = -1 - (-1) - (-1) = 1$$

$X = \{b\}$ $F = \{a, b\}$

$$\mu_\varrho(X, F) = -\sum_{X \subseteq z \subset F} \mu_\varrho(X, z) = -\mu_\varrho(\{b\}, \{b\}) = -1$$

$X = \{b\}$ $F = \{b, c, d\}$

$$\mu_\varrho(X, F) = -\sum_{X \subseteq z \subset F} \mu_\varrho(X, z) = -\mu_\varrho(\{b\}, \{b\}) = -1$$

$X = \{b\}$ $F = \{a, b, c, d\}$

$$\mu_\varrho(X, F) = -\sum_{X \subseteq z \subset F} \mu_\varrho(X, z)$$
$$= -\mu_\varrho(\{b\}, \{b\}) - \mu_\varrho(\{b\}, \{a, b\})$$
$$- \mu_\varrho(\{b\}, \{b, c, d\}) = -1 - (-1) - (-1) = 1$$

$X = \{c, d\}$ $F = \{a, c, d\}$

$$\mu_\varrho(X, F) = -\sum_{X \subseteq z \subset F} \mu_\varrho(X, z) = -\mu_\varrho(\{c, d\}, \{c, d\}) = -1$$

$X = \{c, d\}$ $F = \{b, c, d\}$

$$\mu_\varrho(X, F) = -\sum_{X \subseteq z \subset F} \mu_\varrho(X, z) = -\mu_\varrho(\{c, d\}, \{c, d\}) = -1$$

$X = \{c, d\}$ $F = \{a, b, c, d\}$

$$\mu_{\mathfrak{L}}(X, F) = -\sum_{X \subseteq z \subset F} \mu_{\mathfrak{L}}(X, z)$$
$$= -\mu_{\mathfrak{L}}(\{c, d\}, \{c, d\}) - \mu_{\mathfrak{L}}(\{c, d\}, \{a, b, c, d\})$$
$$- \mu_{\mathfrak{L}}(\{c, d\}, \{a, b, c, d\}) = -1 - (-1) - (-1) = 1$$

$X = \{a, b\}$ $F = \{a, b, c, d\}$

$$\mu_{\mathfrak{L}}(X, F) = -\sum_{X \subseteq z \subset F} \mu_{\mathfrak{L}}(X, z) = -\mu_{\mathfrak{L}}(\{a, b\}, \{a, b\}) = -1$$

$X = \{a, c, d\}$ $F = \{a, b, c, d\}$

$$\mu_{\mathfrak{L}}(X, F) = -\sum_{X \subseteq z \subset F} \mu_{\mathfrak{L}}(X, z) = -\mu_{\mathfrak{L}}(\{a, c, d\}, \{a, c, d\}) = -1$$

$X = \{a, b\}$ $F = \{a, b, c, d\}$

$$\mu_{\mathfrak{L}}(X, F) = -\sum_{X \subseteq z \subset F} \mu_{\mathfrak{L}}(X, z) = -\mu_{\mathfrak{L}}(\{b, c, d\}, \{b, c, d\}) = -1$$

The table below presents the values for the Möbius function for the matroid:

$X \backslash F$	$\emptyset$	$\{a\}$	$\{b\}$	$\{c, d\}$	$\{a, b\}$	$\{a, c, d\}$	$\{b, c, d\}$	$\{a, b, c, d\}$
$\emptyset$	1	−1	−1	−1	1	1	1	−1
$\{a\}$	0	1	0	0	−1	−1	0	1
$\{b\}$	0	0	1	0	−1	0	−1	1
$\{c, d\}$	0	0	0	1	0	−1	−1	1
$\{a, b\}$	0	0	0	0	1	0	0	−1
$\{a, c, d\}$	0	0	0	0	0	1	0	−1
$\{b, c, d\}$	0	0	0	0	0	0	1	−1
$\{a, b, c, d\}$	0	0	0	0	0	0	0	1

5.6.6   DEFINITION: MÖBIUS INVARIANT

Let M be a matroid with ground set S and lattice of flats $\mathfrak{L}$. The Möbius invariant of M is $\mu_M(\emptyset, S) = \mu_{\mathfrak{L}}(\emptyset, S)$.

5.6.7   EXAMPLE

Compute the Möbius invariant for the matroids in examples 5.2.3-5.

In each of these examples we computed a table of the Möbius function. All we need do is look up the appropriate entry on each table.

5.2.3   In this case $\emptyset \notin \mathfrak{L}$, so $\mu_M(\emptyset, S) = \mu_{\mathfrak{L}}(\emptyset, S) = 0$

5.2.4 $\mu_M(\emptyset, S) = \mu_{\mathfrak{L}}(\emptyset, S) = 2$

5.2.5 $\mu_M(\emptyset, S) = \mu_{\mathfrak{L}}(\emptyset, \{a, b, c, d\}) = -1$

5.6.8   MÖBIUS INVARIANT PROPERTIES

We state without proof a few important properties of the Möbius invariant. In the following, M is a matroid with ground set S and lattice of flats $\mathfrak{L}$; M_1 and M_2 are matroids where $M = M_1 \oplus M_2$; e is an element of M that is neither an isthmus nor a loop; and $B_i \in \mathfrak{L}$, $B_i \subset B$ such that $e \in B$ but $e \notin B_i$. Finally, in expressions below, $\widehat{\emptyset} = cl(\emptyset)$.

MI-1 $\mu_M(A, B) = \sum_{\substack{A \subseteq X \subseteq B \\ cl(X) = B}} (-1)^{|X - A|}$

MI-2 $\mu_M(\widehat{\emptyset}, B) = -\sum_i \mu_M(\widehat{\emptyset}, B_i)$

MI-3 $\mu_M(\widehat{\emptyset}, S) = \mu_{M \setminus e}(\widehat{\emptyset}, S - e) - \mu_{M/e}(\widehat{\emptyset}, S - e)$

MI-4 $\mu_M(\widehat{\emptyset}, S) = \mu_{M_1}(\widehat{\emptyset}, S_1) \mu_{M_2}(\widehat{\emptyset}, S_2)$

MI-5 $\left| \mu_M(\widehat{\emptyset}, S) \right| = \left| \mu_{M \setminus e}(\widehat{\emptyset}, S - e) \right| + \left| \mu_{M/e}(\widehat{\emptyset}, S - e) \right|$

MI-6 $\left| \mu_M(\widehat{\emptyset}, S) \right| = \left| \mu_{M_1}(\widehat{\emptyset}, S_1) \right| \left| \mu_{M_2}(\widehat{\emptyset}, S_2) \right|$

5.7   Characteristic Polynomial

The characteristic polynomial of a matroid is an extension of the chromatic polynomial of a graph. The characteristic polynomial is defined in terms of the Möbius function for a matroid.

5.7.1   DEFINITION: CHARACTERISTIC POLYNOMIAL

Let M be a matroid with rank function r, lattice of flats $\mathfrak{L}$, and Möbius function μ_M. The characteristic polynomial of M is a polynomial in λ defined as

$$\chi_M(\lambda) = \sum_{F \in \mathfrak{L}} \mu_M(\hat{0}, F)\lambda^{r(M)-r(F)}$$

5.7.2   EXAMPLE

Compute the characteristic polynomial for the matroid with ground set and basis given by

$$S = \{a, b, c\}$$
$$B = \{\{a, b\}, \{a, b\}, \{b, c\}\}^{\cdot}$$

This is the matroid of example 5.2.5. The lattice of flats is

$$
\begin{array}{cl}
0 & \varnothing \\
1 & \{a\}\{b\}\{c\} \\
2 & \{a, b, c\}
\end{array}
$$

and Möbius function

$X \backslash F$	$\varnothing$	$\{a\}$	$\{b\}$	$\{c\}$	$\{a, b, c\}$
$\varnothing$	1	-1	-1	-1	2
$\{a\}$	0	1	0	0	-1
$\{b\}$	0	0	1	0	-1
$\{c\}$	0	0	0	1	-1
$\{a, b, c\}$	0	0	0	0	1

From the specification of the Möbius function we can compute the characteristic polynomial.

$$\chi_M(\lambda) = \sum_{F \in \mathfrak{L}} \mu_M(\varnothing, F)\lambda^{r(M)-r(F)}$$

$$= \mu_M(\varnothing, \varnothing)\lambda^{2-r(\varnothing)} + \mu_M(\varnothing, \{a\})\lambda^{2-r(\{a\})} + \mu_M(\varnothing, \{b\})\lambda^{2-r(\{b\})}$$
$$+ \mu_M(\varnothing, \{c\})\lambda^{2-r(\{c\})} + \mu_M(\varnothing, \{a, b, c\})\lambda^{2-r(\{a,b,c\})}$$

$$= 1 \cdot \lambda^{2-0} + (-1) \cdot \lambda^{2-1} + (-1) \cdot \lambda^{2-1} + (-1) \cdot \lambda^{2-1} + 2 \cdot \lambda^{2-2}$$

$$= \lambda^2 - 3\lambda + 2$$

5.7.3 EXAMPLE

Compute the characteristic polynomial for the matroid with ground set and basis given by

$$S = \{a, b, c, d\}$$
$$B = \{\{a, b, c\}, \{a, b, d\}\}$$

This is the matroid of example 5.2.6. The lattice of flats is

0	$\emptyset$
1	$\{a\}\{b\}\{c, d\}$
2	$\{a, b\}\{a, c, d\}\{b, c, d\}$
3	$\{a, b, c, d\}$

and Möbius function

$X\backslash F$	$\emptyset$	$\{a\}$	$\{b\}$	$\{c, d\}$	$\{a, b\}$	$\{a, c, d\}$	$\{b, c, d\}$	$\{a, b, c, d\}$
$\emptyset$	1	−1	−1	−1	1	1	1	−1
$\{a\}$	0	1	0	0	−1	−1	0	1
$\{b\}$	0	0	1	0	−1	0	−1	1
$\{c, d\}$	0	0	0	1	0	−1	−1	1
$\{a, b\}$	0	0	0	0	1	0	0	−1
$\{a, c, d\}$	0	0	0	0	0	1	0	−1
$\{b, c, d\}$	0	0	0	0	0	0	1	−1
$\{a, b, c, d\}$	0	0	0	0	0	0	0	1

We compute the characteristic polynomial similar to the previous example.

$$\chi_M(\lambda) = \sum_{F \in \mathfrak{L}} \mu_M(\emptyset, F)\lambda^{r(M)-r(F)}$$

$$= \mu_M(\emptyset, \emptyset)\lambda^{3-r(\emptyset)} + \mu_M(\emptyset, \{a\})\lambda^{3-r(\{a\})} + \mu_M(\emptyset, \{b\})\lambda^{3-r(\{b\})}$$
$$+ \mu_M(\emptyset, \{c, d\})\lambda^{3-r(\{c,d\})} + \mu_M(\emptyset, \{a, b\})\lambda^{3-r(\{a,b\})}$$
$$+ \mu_M(\emptyset, \{a, b, c\})\lambda^{3-r(\{a,b,c\})}$$
$$+ \mu_M(\emptyset, \{b, c, d\})\lambda^{3-r(\{b,c,d\})}$$
$$+ \mu_M(\emptyset, \{a, b, c, d\})\lambda^{3-r(\{a,b,c,d\})}$$

$$= 1 \cdot \lambda^{3-0} + (-1) \cdot \lambda^{3-1} + (-1) \cdot \lambda^{3-1} + (-1) \cdot \lambda^{3-1} + 1 \cdot \lambda^{3-2}$$
$$+ 1 \cdot \lambda^{3-2} + 1 \cdot \lambda^{3-2} + (-1) \cdot \lambda^{3-3}$$

$$= \lambda^3 - 3\lambda^2 + 3\lambda - 1$$

5.7.4 PROPOSITION

Let M be a matroid with ground set S, rank function r, and lattice of flats $\mathfrak{L}$. The characteristic polynomial of M is

CP-1 $\chi_M(\lambda) = \sum_{X \subseteq S} (-1)^{|X|} \lambda^{r(M) - r(X)}$

Proof. This result follows from **MI-1**. We have

$$\mu_M(\hat{\emptyset}, F) = \sum_{\substack{\emptyset \subseteq X \subseteq F \\ cl(X) = F}} (-1)^{|X|}$$

Substituting this into the expression for the characteristic polynomial,

$$\chi_M(\lambda) = \sum_{F \in \mathfrak{L}} \sum_{\substack{\emptyset \subseteq X \subseteq F \\ cl(X) = F}} (-1)^{|X|} \lambda^{r(M) - r(F)}$$

We know that the closure of every subset of S is a unique flat $F \in \mathfrak{L}$. Thus, the double sum is actually summing over every subset of S because each subset of S has a closure which is a flat.

Also, we know $r(X) = r\big(cl(X)\big) = r(F)$, where F is the flat resulting from the closure of X.

Replacing the double sum with a single sum over all subsets of S,

$$\chi_M(\lambda) = \sum_{X \subseteq S} (-1)^{|X|} \lambda^{r(M) - r(X)} \quad \blacksquare$$

5.7.5 PROPOSITION

Let M be a matroid with ground set S, and lattice of flats $\mathfrak{L}$. Let $e \in S$ be an element of the ground set. The characteristic polynomial of M satisfies

$$\chi_M(\lambda) = \chi_{M \setminus e}(\lambda) - \chi_{M/e}(\lambda)$$

Proof. This result follows from **MI-3**. $\blacksquare$

5.7.6   PROPOSITION

Let M be a matroid with ground set S, and lattice of flats $\mathcal{L}$. Let $e \in S$ be an element of the ground set. The characteristic polynomial of M satisfies

$$\chi_M(\lambda) = \chi_{M \setminus e}(\lambda)\chi_{M/e}(\lambda)$$

Proof. This result follows from **MI-4**. ∎

5.7.7   DEFINITION: BETA INVARIANT

> Let M be a matroid with rank function r, lattice of flats $\mathcal{L}$, and characteristic polynomial $\chi_M(\lambda)$. The beta invariant of M is defined as
>
> $$\beta_M = (-1)^{r(M)-1}\frac{d}{d\lambda}\chi_M(\lambda)\Big|_{\lambda=1}$$

The beta invariant may be written in terms of the Möbius function by substituting the expression for the characteristic polynomial into the definition

$$\beta_M = (-1)^{r(M)-1}\frac{d}{d\lambda}\sum_{F \in \mathcal{L}} \mu_M(\hat{0}, F)\lambda^{r(M)-r(F)}\Big|_{\lambda=1}$$

$$= (-1)^{r(M)-1}\sum_{F \in \mathcal{L}} \mu_M(\hat{0}, F)[r(M) - r(F)]$$

$$= (-1)^{r(M)-1}r(M)\sum_{F \in \mathcal{L}} \mu_M(\hat{0}, F) - (-1)^{r(M)-1}\sum_{F \in \mathcal{L}} \mu_M(\hat{0}, F)r(F)$$

We know $\sum_{F \in \mathcal{L}} \mu_M(\emptyset, F) = 0$ from the definition of the Möbius function on a matroid. Specifically, if $A \subset B$,

$$\mu_M(A, B) = -\sum_{\substack{A \subseteq z \subset B \\ z \in \mathcal{L}}} \mu_M(A, z)$$

so

$$\mu_M(A, B) - \sum_{\substack{A \subseteq z \subset B \\ z \in \mathfrak{L}}} \mu_M(A, z) = \sum_{\substack{A \subseteq z \subseteq B \\ z \in \mathfrak{L}}} \mu_M(A, z) = 0$$

Thus,

$$\sum_{F \in \mathfrak{L}} \mu_M(\hat{\emptyset}, F) = \sum_{\substack{\emptyset \subseteq z \subseteq S \\ z \in \mathfrak{L}}} \mu_M(\hat{\emptyset}, z) = 0$$

This leaves us with

$$\beta_M = (-1)^{r(M)} \sum_{F \in \mathfrak{L}} \mu_M(\hat{\emptyset}, F) \, r(F)$$

5.7.8 EXAMPLE

Compute the beta invariant for the matroid in example 5.4.2.

The rank of this matroid is $r(M) = 2$, and the characteristic polynomial is

$$\chi_M(\lambda) = \lambda^2 - 3\lambda + 2$$

Computing the beta invariant,

$$\beta_M = (-1)^{r(M)-1} \frac{d}{d\lambda} \chi_M(\lambda) \Big|_{\lambda=1} = (-1)^{2-1} [2\lambda - 3]_{\lambda=1}$$

$$= (-1)(-1) = 1$$

5.7.9 EXAMPLE

Compute the beta invariant for the matroid in example 5.4.3.

The rank of this matroid is $r(M) = 3$, and the characteristic polynomial is

$$\chi_M(\lambda) = \lambda^3 - 3\lambda^2 + 3\lambda - 1$$

Computing the beta invariant,

$$\beta_M = (-1)^{r(M)-1} \frac{d}{d\lambda} \chi_M(\lambda) \Big|_{\lambda=1} = (-1)^{3-1} [4\lambda^2 - 6\lambda + 3]_{\lambda=1}$$

$$= (-1)^2 (1) = 1$$

5.7.10 PROBLEMS

1. Let $M = (S, B)$ be the matroid
$$S = \{a, b, c\}$$
$$B = \{\{a, c\}, \{b, c\}\}$$
   a. Find the characteristic polynomial of M.
   b. Find the beta invariant.

5.8   Whitney Numbers

Whitney numbers arise from the coefficients of the characteristic and rank polynomials of a matroid. The Whitney numbers appear in many contexts and it is useful to review their properties for later use.

5.8.1   DEFINITION: WHITNEY NUMBERS OF THE FIRST KIND

Let M be a matroid with rank function r, ground set S, lattice of flats $\mathfrak{L}$, and characteristic polynomial $\chi_M(\lambda)$. The Whitney numbers of the first kind are the coefficients of the characteristic polynomial

$$\chi_M(\lambda) = \sum_{F \in \mathfrak{L}} \mu_M(\emptyset, F) \lambda^{r(M)-r(F)} = \sum_{n=0}^{r(M)} w_n \lambda^n$$

It is useful to have an unsigned version of these numbers. Let
$$w_n^+ = |w_n|$$
We call these the unsigned Whitney numbers of the first kind. The Whitney numbers of the first kind alternate in sign. We will be interested in the growth properties of the Whitney numbers. As such, the sign is of little value, only the magnitude is needed.

5.8.2   DEFINITION: WHITNEY NUMBERS OF THE SECOND KIND

Let M be a matroid with rank function r, ground set S, lattice of flats $\mathfrak{L}$, and rank polynomial $\rho_M(\lambda)$. The Whitney numbers of the second kind are the coefficients of the characteristic polynomial

$$\rho_M(\lambda) = \sum_{F \in \mathfrak{L}} \lambda^{r(M)-r(F)} = \sum_{n=0}^{r(M)} W_n \lambda^n$$

5.8.3 EXAMPLE

Compute the Whitney numbers for the matroid with ground set and basis given by

$$S = \{a, b, c\}$$
$$B = \{\{a, b\}, \{a, b\}, \{b, c\}\}^{\cdot}$$

The characteristic and rank polynomials are found in example 5.7.2 and 5.4.2 respectively.

$$\chi_M(\lambda) = \lambda^2 - 3\lambda + 2$$
$$\rho_M(\lambda) = \lambda^2 + 3\lambda + 1$$

From these expressions we can read off the Whitney numbers.

$$w_0 = 2 \quad w_1 = -3 \quad w_2 = 1$$

$$W_0 = 1 \quad W_1 = 3 \quad W_2 = 1$$

5.8.4 EXAMPLE

Compute the Whitney numbers for the matroid with ground set and basis given by

$$S = \{a, b, c, d\}$$
$$B = \{\{a, b, c\}, \{a, b, d\}\}^{\cdot}$$

The characteristic and rank polynomials are found in example 5.7.3 and 5.4.3 respectively.

$$\chi_M(\lambda) = \lambda^3 - 3\lambda^2 + 3\lambda - 1$$
$$\rho_M(\lambda) = \lambda^3 + 3\lambda^2 + 3\lambda + 1$$

From these expressions we can read off the Whitney numbers.

$$w_0 = -1 \quad w_1 = 3 \quad w_2 = -3 \quad w_3 = 1$$

$$W_0 = 1 \quad W_1 = 3 \quad W_2 = 3 \quad W_3 = 1$$

5.8.5 PROPERTIES

Here we list some properties of the Whitney numbers. In the following, M is a matroid with rank $r \geq 3$ on a ground set S where $n = |S|$.

WN-1 $\binom{r}{k} + (n-r)\binom{r-1}{k-1} \leq w_k^+ \leq \binom{n}{k}$

WN-2 $\binom{r}{k} + (n-r)\binom{r-2}{k-1} \leq W_k^+ \leq \binom{n}{k}$

5.9 Tutte-Grothendieck Invariant

The Tutte-Grothendieck (TG) invariant is a general matroid invariant that has a wide range of applications. In fact, the TG matroid invariant is a generalization of the Tutte polynomial, which is itself a generalization of the chromatic polynomial.

5.9.1 DEFINITION: TUTTE-GROTHENDIECK INVARIANT

> Let $\mathfrak{M}$ be a class of matroids that are closed under isomorphism. Let f be a map $f: \mathfrak{M} \to R$ where R is a ring, and let $M \in \mathfrak{M}$ be a matroid, and $e \in M$ be an element of the ground set of M. f is a TG invariant if f satisfies
>
> **TG-1** $f(M) = \sigma f(M - e) + \tau f(M/e)$ if e is not an isthmus or loop
>
> **TG-2** $f(M) = f(M|e)f(M/e)$ is if e is an isthmus or loop

This definition is a recursion relation for f. The invariant will present itself as a polynomial in σ and τ. We show that the rank generating function and the Tutte polynomial are both TG invariants. In fact, they are related by a linear transformation of variables.

Note we use the notation $M - e$ instead of $M \backslash e$ in the above definition. These are equivalent, however, we choose to use $M - e$ to expressly make the point that the ground set of the matroid $M - e$ is simply the ground set of M with the element e removed.

5.9.2 THEOREM

Let M be a matroid with rank function r, ground set S, lattice of flats $\mathfrak{L}$, and rank generating function $R_M(\theta, \phi)$. The rank generating function is a TG invariant.

Proof. The rank generating function of M is a polynomial in θ and ϕ defined as

$$R_M(\theta, \phi) = \sum_{X \subseteq S} \theta^{r(M)-r(X)} \phi^{|X|-r(X)}$$

First, choose some $e \in S$ and split the sum into components where $e \in X$ and $e \notin X$.

$$R_M(\theta, \phi) = \sum_{\substack{X \subseteq S \\ e \notin X}} \theta^{r(M)-r(X)} \phi^{|X|-r(X)} + \sum_{\substack{X \subseteq S \\ e \in X}} \theta^{r(M)-r(X)} \phi^{|X|-r(X)}$$

Examine the first sum, and look to link the sum to the rank function on $M - e$. Form the restriction matroid $M|(M - e)$. Examine the set of bases B_M as compared to B_{M-e}. All of the bases in B_M that do not contain the element e will be present in B_{M-e}.

If there are some bases in B_M that do not contain e, then $r(M) = r(M - e)$ because the rank is the cardinality of any basis, and all basis have the same cardinality. So if any basis from B_M is present in B_{M-e}, then the matroids have the same rank.

If every basis in B_M contains the element e, then e is an isthmus (by definition). In this case, $r(M) = r(M - e) + 1$ because none of the bases B_M will be present in B_{M-e}, but at least one basis will be present with cardinality one $r(M) - 1$. We know that this must be the case because all subsets of a basis are independent sets. Therefore, there is a subset of a basis that does not contain e (form the subset by taking any baisis, then remove e). This independent set will be a basis of B_{M-e}, and it has dimension $r(M) - 1$.

From these two results we know

$$r(M) = \begin{cases} r(M - e) & e \text{ is not an isthmus} \\ r(M - e) + 1 & e \text{ is an isthmus} \end{cases}$$

We can use this result to eliminate $r(M)$ in the first sum. Moreover, we recognize that the sum $X \subseteq S$ such that $e \notin X$ is equivalent to the sum over $X \subseteq S - e$.

$$\sum_{\substack{X \subseteq S \\ e \notin X}} \theta^{r(M)-r(X)} \phi^{|X|-r(X)} = \sum_{X \subseteq S-e} \theta^{r(M)-r(X)} \phi^{|X|-r(X)}$$

$$
= \begin{cases} \displaystyle\sum_{X \subseteq S-e} \theta^{r(M-e)-r(X)} \phi^{|X|-r(M-e)} & e \text{ is not an isthmus} \\[4mm] \displaystyle\sum_{X \subseteq S-e} \theta^{r(M-e)+1-r(X)} \phi^{|X|-r(X)} & e \text{ is an isthmus} \end{cases}
$$

$$
= \begin{cases} \displaystyle\sum_{X \subseteq S-e} \theta^{r(M-e)-r(X)} \phi^{|X|-r(M-e)} & e \text{ is not an isthmus} \\[4mm] \theta\displaystyle\sum_{X \subseteq S-e} \theta^{r(M-e)-r(X)} \phi^{|X|-r(X)} & e \text{ is an isthmus} \end{cases}
$$

$$
= \begin{cases} R_{M-e}(\theta,\phi) & e \text{ is not an isthmus} \\ \theta R_{M-e}(\theta,\phi) & e \text{ is an isthmus} \end{cases}
$$

Next, examine the second sum. We know that every X in this sum contains the element e. We can write this sum in a similar form to the first sum as

$$
\sum_{\substack{X \subseteq S \\ e \in X}} \theta^{r(M)-r(X)} \phi^{|X|-r(X)} = \sum_{Y \subseteq S-e} \theta^{r([M-e] \cup e)-r(Y \cup e)} \phi^{|Y \cup e|-r(Y \cup e)}
$$

Now, examine how the rank function on the contraction $r_{M/e}(Y)$ compares to $r(Y \cup e)$. We know from **CP-1** that the rank function on the contraction obeys

$$
r_{M/e}(X) = r_M(X \cup T) - r_M(T).
$$

Set $X = Y$ and $T = e$:

$$
r_{M/e}(Y) = \begin{cases} r(Y \cup e) - 1 & e \text{ is not a loop} \\ r(Y \cup e) & e \text{ is a loop} \end{cases}
$$

Furthermore, the sum $|Y \cup e| - r(Y \cup e)$ becomes

$$
|Y| - r_{M/e}(Y) = \begin{cases} |Y \cup e| - r(Y \cup e) & e \text{ is not a loop} \\ |Y \cup e| - r(Y \cup e) - 1 & e \text{ is a loop} \end{cases}
$$

Finally, remember that

$$
r_{M/e}(M - e) = \begin{cases} r([M - e] \cup e) - 1 & e \text{ is not a loop} \\ r([M - e] \cup e) & e \text{ is a loop} \end{cases}
$$

Substituting these into the second sum,

$$\sum_{Y \subseteq S - e} \theta^{r([M-e] \cup e) - r(Y \cup e)} \phi^{|Y \cup e| - r(Y \cup e)}$$

$$= \begin{cases} \displaystyle\sum_{Y \subseteq S - e} \theta^{r_{M/e}([M-e]) + 1 - r_{M/e}(Y) - 1} \phi^{|Y| - r_{M/e}(Y)} & e \text{ is not a loop} \\[2em] \displaystyle\sum_{Y \subseteq S - e} \theta^{r_{M/e}([M-e]) - r_{M/e}(Y)} \phi^{|Y| - r_{M/e}(Y) + 1} & e \text{ is a loop} \end{cases}$$

$$= \begin{cases} \displaystyle\sum_{Y \subseteq S - e} \theta^{r_{M/e}([M-e]) - r(Y)} \phi^{|Y| - r_{M/e}(Y)} & e \text{ is not a loop} \\[2em] \phi \displaystyle\sum_{Y \subseteq S - e} \theta^{r_{M/e}([M-e]) - r(Y)} \phi^{|Y| - r_{M/e}(Y)} & e \text{ is a loop} \end{cases}$$

$$= \begin{cases} R_{M/e}(\theta, \phi) & e \text{ is not a loop} \\ \phi R_{M/e}(\theta, \phi) & e \text{ is a loop} \end{cases}$$

Combining the results of the first and second sums,

$$R_M(\theta, \phi) = \begin{cases} R_{M-e}(\theta, \phi) + R_{M/e}(\theta, \phi) & e \text{ is neither a isthmus or loop} \\ \theta R_{M-e}(\theta, \phi) + R_{M/e}(\theta, \phi) & e \text{ is an isthmus} \\ R_{M-e}(\theta, \phi) + \phi R_{M/e}(\theta, \phi) & e \text{ is a loop} \end{cases}$$

From **CP-2**, if e is a loop or isthmus, $M - e = M/e$. With this,

$$R_M(\theta, \phi) = \begin{cases} R_{M-e}(\theta, \phi) + R_{M/e}(\theta, \phi) & e \text{ is neither a isthmus or loop} \\ (\theta + 1) R_{M-e}(\theta, \phi) & e \text{ is an isthmus} \\ (\phi + 1) R_{M/e}(\theta, \phi) & e \text{ is a loop} \end{cases}$$

Finally, if M is a loop then $R_M(\theta, \phi) = \phi + 1$. Similarly, if M is an isthmus, $R_M(\theta, \phi) = \theta + 1$. Thus, when e is a loop, $R_{M|e}(\theta, \phi) = \phi + 1$. Moreover, when e is an isthmus, $R_{M|e}(\theta, \phi) = \theta + 1$.

Substituting into the expression above,

$$R_M(\theta, \phi) = \begin{cases} R_{M-e}(\theta, \phi) + R_{M \setminus e}(\theta, \phi) & e \text{ is neither a isthmus or loop} \\ R_{M|e}(\theta, \phi) R_{M-e}(\theta, \phi) & e \text{ is an isthmus} \\ R_{M|e}(\theta, \phi) R_{M \setminus e}(\theta, \phi) & e \text{ is a loop} \end{cases}$$

We see that the rank function satisfies the properties of a TG invariant. ∎

5.9.3   DEFINITION: TUTTE POLYNOMIAL

> Let $\mathfrak{M}$ be a class of matroids that are closed under isomorphism. Let t be a map $f: \mathfrak{M} \to R$ where R is a ring, and let $M \in \mathfrak{M}$ be a matroid, and $e \in M$ be an element of the ground set of M. t is the Tutte polynomial if t satisfies
>
> **TP-1** $t(M) = t(M - e) + t(M/e)$ if e is not an isthmus or loop
>
> **TP-2** $t(M) = t(M - e)t(M/e)$ is e is an isthmus or loop
>
> **TP-3** $t(M) = x$ is M is an isthmus
>
> **TP-4** $t(M) = y$ is M is a loop

It is easy to see from this definition that the Tutte polynomial is a TG invariant. Furthermore, the Tutte polynomial is equivalent to the rank generating function. If we make the substitutions $x = \theta + 1$, and $y = \phi + 1$, the Tutte polynomial conforms to the rank generating polynomial.

5.10   Distribution Polynomial

The distribution polynomial is a polynomial invariant based on how the elements are distributed in the matroid bases. The distribution polynomial uniquely identifies all matroids with rank 7 or less, meaning that if the distribution polynomial is the same for two matroids with rank 7 or less, then the matroids are in fact the same.

5.10.1   DEFINITION: BASIS ELEMENT SWAP OPERATOR

> Let M be a matroid on a ground set S with basis B. The Swap Operator is a map $\sim: S \times S \to \{0,1\}$ such that
> $$e_i \sim e_j = \begin{cases} 1 & B' = B \\ 0 & B' \neq B \end{cases}$$
> where $e_i, e_j \in S$ and B' is the basis obtained by swapping e_i and e_j in B. From here forward, when using the $\sim$ operator, $e_i \sim e_j$ will be used to mean that $e_i \sim e_j = 1$, while $e_i \nsim e_j$ will be used to mean $e_i \sim e_j = 0$.

The swap operator is the permutation (i, j) acting on the matroid. If the basis is invariant under the permutation, then $e_i \sim e_j$, otherwise $e_i \not\sim e_j = 0$.

5.10.2 THEOREM

The Basis Element Swap Operator forms an equivalence relation.

Proof.

The operator $\sim$ is an equivalence relation iff the operator obeys the three rules:

1. Reflexivity: $a \sim a$
2. Symmetry: $a \sim b \rightarrow b \sim a$
3. Transitivity: $a \sim b, b \sim c \rightarrow a \sim c$

We proceed to show that each of these properties is satisfied by the $\sim$ operator. In the following, let M be a matroid on a ground set S with basis B, and let $a, b, c \in S$.

1. Reflexivity: $a \sim a$

Swapping an element a with itself is realized through the identity permutation. This makes no changes to any of the bases in the basis. Thus, $B' = B$ so $a \sim a$.

2. Symmetry: $a \sim b \rightarrow b \sim a$

Swapping element a with b results in a new basis set B' which we obtain by acting the permutation $(a\, b)$ on the basis set B. If we instead swap b with a, we need to use the permutation $(b\, a)$ on the basis set B.

However, as a permutation, $(a\, b) = (b\, a)$, so $(b\, a)$ acting on B results in B'. Thus, $a \sim b \rightarrow b \sim a$.

2. Transitivity: $a \sim b, b \sim c \rightarrow a \sim c$

In terms of permutations, $a \sim b$ means

$$(a\, b)B = B$$

Similarly,

$$(b\ c)B = B$$

But the permutation $(a\ c) = (a\ b)(b\ c)(a\ b)$. Thus,
$$(a\ c)B = (a\ b)(b\ c)(a\ b)B$$

$$= (a\ b)(b\ c)B$$

$$= (a\ b)B$$

$$= B$$

So if we have $a{\sim}b$ and $b{\sim}c$ then $a{\sim}c$. ∎

5.10.3 DEFINITION: SWAP EQUIVALENCE SET

> Let M be a matroid on a ground set S with basis B. The Swap Equivalence Set is the set of equivalence classes under the $\sim$ operator.

5.10.4 EXAMPLE

Find the Swap Equivalence Set for the matroid

$$S = \{a, b, c, d, e\}$$
$$B = \{\{a, b, c\}, \{a, b, d\}, \{a, c, d\}, \{b, c, d\}\}'$$

We find the equivalence classes by testing pairs of elements.

Check $a{\sim}b$:

Swap a and b in the basis:

$$B' = \{\{b, a, c\}, \{b, a, d\}, \{b, c, d\}, \{a, c, d\}\}$$

The order of elements in a set is immaterial. We can rearrange the elements of B', and we can rearrange the elements of the bases in B' as well. Doing this we have,

$$B' = \{\{a, b, c\}, \{a, b, d\}, \{a, c, d\}, \{b, c, d\}\}$$

$$= B$$

We see that swapping a and b in the basis results in the same presentation of B. Thus, $a \sim b$.

Check $a \sim c$:

Swap a and c in the basis:

$$B' = \{\{c, b, a\}, \{c, b, d\}, \{c, a, d\}, \{b, a, d\}\}$$

$$= \{\{a, b, c\}, \{a, b, d\}, \{a, c, d\}, \{b, c, d\}\}$$

$$= B$$

Swapping a and c in the basis results in the same presentation of B. Thus, $a \sim c$.

Check $a \sim d$:

Swap a and d in the basis:

$$B' = \{\{d, b, c\}, \{d, b, a\}, \{d, c, a\}, \{b, c, a\}\}$$

$$= \{\{a, b, c\}, \{a, b, d\}, \{a, c, d\}, \{b, c, d\}\}$$

$$= B$$

Swapping a and d in the basis results in the same presentation of B. Thus, $a \sim d$.

Check $a \sim e$:

Swap a and e in the basis:

$$B' = \{\{e, b, c\}, \{e, b, d\}, \{e, c, d\}, \{b, c, d\}\}$$

$$= \{\{b, c, e\}, \{b, d, e\}, \{c, d, e\}, \{b, c, d\}\}$$

$$\neq B$$

Swapping a and d in the basis results in a different presentation of B. Thus, $a \not\sim e$.

From the above, there are two equivalence classes: $\{a, b, c, d\}$ and $\{e\}$. The Swap Class is

$$S_C = \{\{a, b, c, d\}, \{e\}\}.$$

5.10.5 EXAMPLE

Find the Swap Equivalence Set for the matroid

$$S = \{a, b, c, d\}$$
$$B = \{\{a, b\}, \{a, c\}, \{a, d\}, \{b, c\}, \{b, d\}\}$$

We find the equivalence classes by testing pairs of elements.

Check $a \sim b$:

Swap a and b in the basis:

$$B' = \{\{b, a\}, \{b, c\}, \{b, d\}, \{a, c\}, \{a, d\}\}$$

$$= \{\{a, b\}, \{a, c\}, \{a, d\}, \{b, c\}, \{b, d\}\}$$

$$= B$$

So, $a \sim b$.

Check $a \sim c$:

Swap a and c in the basis:

$$B' = \{\{c, b\}, \{c, a\}, \{c, d\}, \{b, a\}, \{b, d\}\}$$

$$= \{\{a, b\}, \{a, c\}, \{b, c\}, \{b, d\}, \{c, d\}\}$$

$$\neq B$$

So, $a \not\sim c$.

Check $a \sim d$:

Swap a and d in the basis:

$$B' = \{\{d,b\}, \{d,c\}, \{d,a\}, \{b,c\}, \{b,a\}\}$$

$$= \{\{a,b\}, \{a,d\}, \{b,c\}, \{b,d\}, \{c,d\}\}$$

$$\neq B$$

So, $a \nsim d$.

From this, we see that $\{a,b\}$ must be an equivalence class.

Check $c\sim d$:

Swap c and d in the basis:

$$B' = \{\{a,b\}, \{a,d\}, \{a,c\}, \{b,d\}, \{b,c\}\}$$

$$= \{\{a,b\}, \{a,c\}, \{a,d\}, \{b,c\}, \{b,d\}\} =$$

$$\neq B$$

Swapping b and c in the basis results in a different presentation of B. Thus, $c\sim d$.

The Swap Equivalence Set is

$$S_C = \{\{a,b\}, \{c,d\}\}.$$

5.10.6 DEFINITION: SWAP EQUIVALENCE SET DEGREE

> Let M be a matroid on a ground set S with basis B. The Swap Equivalence Set is the set of equivalence classes under the $\sim$ operator. The degree of the Swap Equivalence Set is the count of the number of bases each element in the equivalence class appears in the basis.

All elements in an equivalence class under the $\sim$ must appear in the same number of bases.

5.10.7 EXAMPLE

Find the degree of each element of Swap Equivalence Set for the matroid

$$S = \{a, b, c, d, e\}$$
$$B = \{\{a, b, c\}, \{a, b, d\}, \{a, c, d\}, \{b, c, d\}\}^{\cdot}$$

The Swap Equivalence Class for this matroid is computed in example 5.10.4:

$$S_C = \{\{a, b, c, d\}, \{e\}\}$$

In the first element, $\{a, b, c, d\}$, each element appears in 3 bases. For example, the element a appears in $\{a, b, c\}, \{a, b, d\}, \{a, c, d\}$, while b appears in $\{a, b, c\}, \{a, b, d\}, \{b, c, d\}$. Consequently, the degree for $\{a, b, c, d\}$ is 3.

Similarly, examining the second element $\{e\}$, e does not appear in any base, so the degree is 0.

Element	Degree
$\{a, b, c, d\}$	3
$\{e\}$	0

5.10.8 EXAMPLE

Find the degree of each element of Swap Equivalence Set for the matroid

$$S = \{a, b, c, d\}$$
$$B = \{\{a, b\}, \{a, c\}, \{a, d\}, \{b, c\}, \{b, d\}\}^{\cdot}$$

The Swap Equivalence Class for this matroid is computed in example5.10.5:

$$S_C = \{\{a, b\}, \{c, d\}\}$$

The degree for each of these is:

Element	Degree
$\{a, b\}$	3
$\{c, d\}$	2

5.10.9 DEFINITION: DISTRIBUTION POLYNOMIAL

Let M be a matroid on a ground set S with basis B, and let S_C be the Swap Equivalence Set for M under the $\sim$ operator. The Distribution Polynomial is the polynomial in two variables as

$$D(\theta, \varphi) = \sum_{s \in S_G} \theta^{d(s)} \varphi^{|s|}$$

where $d(s)$ is the degree of the equivalence class.

5.10.10 EXAMPLE

Find the distribution polynomial for the matroid:

$$S = \{a, b, c, d, e\}$$
$$B = \{\{a, b, c\}, \{a, b, d\}, \{a, c, d\}, \{b, c, d\}\}$$

The degree for each Swap Equivalence Set is computed in example 5.10.7. The distribution polynomial is

$$D(\theta, \varphi) = \theta^{d(\{a,b,c,d\})} \varphi^{|\{a,b,c,d\}|} + \theta^{d(\{e\})} \varphi^{|\{e\}|}$$

$$= \theta^3 \varphi^4 + \theta^0 \varphi^1$$

$$= \theta^3 \varphi^4 + \varphi$$

5.10.11 EXAMPLE

Find the distribution polynomial for the matroid:

$$S = \{a, b, c, d\}$$
$$B = \{\{a, b\}, \{a, c\}, \{a, d\}, \{b, d\}\}$$

The degree for each Swap Equivalence Set is computed in example5.10.8. The distribution polynomial is

$$D(\theta, \varphi) = \theta^{d(\{a,b\})} \varphi^{|\{a,b\}|} + \theta^{d(\{c,d\})} \varphi^{|\{c,d\}|}$$

$$= \theta^3 \varphi^2 + \theta^2 \varphi^2$$

Chapter 6: Algorithms

This chapter examines matroid related algorithms. Matroids satisfy the Greedy Algorithm, where one selects an optimal move at every step to reach an optimal solution. In fact, we show that matroids are the *only* objects that reach an optimal result under the Greedy Algorithm.

6.1   Greedy Algorithm

The greedy algorithm is an optimization algorithm that attempts to find an optimal solution in a series of steps. At each step, the greedy algorithm makes the most optimal choice, and the hope is this will produce a globally optimal solution.

When applied to matroids, the greedy algorithm does produce a globally optimal solution. In fact, the greedy algorithm is guaranteed to find an optimal solution only in matroids.

6.1.1   DEFINITION: GREEDY ALGORITHM

Let S be a finite set. Let T be a collection of subsets of S with the following properties:

GA-1 $\emptyset \in T$

GA-2   If $t \in T$ and $\bar{t} \subseteq t$, then $\bar{t} \in T$

Let w be a map from $w \colon S \to \mathbb{R}$. Define the weight Ω of a subset $X \subseteq S$ as

$$\Omega(X) = \sum_{x \in X} w(x).$$

We desire to find a maximal member of T with maximal (minimal) weight. A greedy algorithm proceeds as follows:

1. Start with $X_0 = \emptyset$
2. Find the set N of all $s \in S - X_n$ such that $s \cup X_n \in T$.
    a. If $N \neq \emptyset$, choose $n \in N$ with $w(n)$ maximum (minimum). Set $X_{n+1} = s \cup X_n$. Repeat step 2.
    b. If $N = \emptyset$, then X_n is the desired optimal set.

The greedy algorithm chooses the optimal value at every step. In general, there is no guarantee that this process will lead to a global maximum. However, we will show that if (S,T) form a matroid, then the greedy algorithm does in fact find a global maximum.

Also, it is important to note that the greedy algorithm will always find a maximal element. The algorithm is designed to find a maximal element with maximum weight. It is possible to have non-maximal element with higher weight if the weight function has negative values.

In the above definition, the weight functions $\Omega(X)$ and $w(x)$ represent the same function, only applied to different constructs. $\Omega(X)$ is the weight function applied to the set $X \subseteq S$, while $w(x)$ is the weight function as applied to an element $x \in X$.

6.1.2   LEMMA

If (S,T) is a matroid, then the optimal solution obtained by the greedy algorithm is a basis of (S,T).

Proof. Recall that all bases of a matroid have the same dimension which is the rank of the matroid. Let M be the matroid (S,T) and let $r(M)$ be the rank of the matroid.

Assume v is the optimal solution obtained from the greedy algorithm. If $|v| = r(M)$, then v is a basis of M and the lemma is proved. Otherwise, it must be the case that $|v| \neq r(M)$. We proceed to show that this cannot be the case.

It is impossible for $|v| > r(M)$, because then we would have an independent set of M with dimension larger than the rank of M, and the rank of M is the dimension of the maximal independent sets.

In the case $|v| < r(M)$, we would have an independent set with dimension less than the rank of M. The set v is not a maximal independent set of M because all maximal independent sets of M have $|b| = r(M)$. Thus, there is some set $b \in T$ such that $v \subset b$.

However, since the greedy algorithm arrives at v as a potential solution, then in step 2 we find there is still some $s \in S - v$ such that $s \cup v \in T$ (specifically, we know some element of b must satisfy this because v is a proper subset of b and **SM-3** of matroids tells us that such an element exists). In this case, the greedy algorithm would add this element and continue. Thus, v cannot be a solution obtained from the greedy algorithm.

Therefore, if v is a solution obtained by the greedy algorithm, then $|v| = r(M)$, making v a basis of M.■

6.1.3 THEOREM

The greedy algorithm obtains a maximal element of maximum weight if (S, T) is a matroid.

Proof. If (S, T) is a matroid, then the solution from the greedy algorithm is a basis for the matroid. We proceed to show that this solution is the maximal element of maximum weight.

First, because the greedy algorithm continues to add elements as long as it is possible to do so, the final solution must be a maximal element.

Let v be the solution from the greedy algorithm and let g be the maximum weight for all subsets of $t \in T$ with some specific dimension $|t| = r$. Order the elements of v and g by decreasing weight. Specifically, $v = \{v_1, v_2, ..., v_r\}$ where $w(v_1) > w(v_2) > \cdots > w(v_r)$, and $g = \{g_1, g_2, ..., g_r\}$ where $w(g_1) > w(g_2) > \cdots > w(g_r)$.

If g is the true optimum and v is not, then we must have $\Omega(g) > \Omega(v)$. If this is the case, then there must be some k such that $w(v_k) < w(g_k)$ (otherwise the weight of v would be at least as big as the weight of g).

Let k be first element such that $w(v_k) < w(g_k)$. Let $u = \{v_1, v_2, ..., v_{k-1}\}$ and $f = \{g_1, g_2, ..., g_k\}$. Because (S, T) is a matroid, from **SM-3** we know that there must be some element $\bar{g} \in f$ such that $u \cup \bar{g} \in T$. However, we know that $w(g_1) > w(g_2) > \cdots > w(g_k) > w(v_k)$.

Given this, the greedy algorithm will choose any of the elements of f before choosing v_k. Furthermore, from **SM-3**, we know that some element $\bar{g} \in f$ satisfies $u \cup \bar{g} \in T$. So the greedy algorithm will not choose v_k, contradicting that v is a solution from the greedy algorithm.

It must be the case that there is no such k satisfying $w(v_k) < w(g_k)$. Therefore, we must have $w(v_k) \geq w(g_k)$ for every k. Given this, the weight $\Omega(v) \geq \Omega(g)$. Since g is the global optimum among maximal elements, no other set can have higher weight. Thus, equality must hold yielding $\Omega(v) = \Omega(g)$. If the weight of v is the same as the weight of g, then either $v = g$, or both v and g are global optima. In either case, the greedy algorithm obtains a maximal element of maximum weight solution if (S, T) is a matroid. ∎

6.1.4   THEOREM

Let S be a set and T a collection of subsets of S. If the system (S, T) obeys

GA-1 $\emptyset \in T$

GA-2   If $t \in T$ and $\bar{t} \subseteq t$, then $\bar{t} \in T$

GA-3   For any weight function $w: S \to \mathbb{R}$ the greedy algorithm produces a maximal element $t \in T$ with maximum weight

then (S, T) is a matroid.

Proof. We need to show that T obeys **SM-1**, **SM-2**, and **SM-3**. It is transparent that **GA-1** is the same as **SM-1**, and **GA-2** is the same as **SM-2**. Thus, we only need to show that **GA-3** implies **SM-3**.

We proceed by assuming **SM-3** does not hold and hope to show that this means that **GA-3** also does not hold. If it is possible that **SM-3** does not hold, then there must be some $A, B \in T$ with $|A| < |B|$, such that there is no element $e \in B - A$ with the property that $A \cup e \in T$. In other words, if during the execution of the greedy algorithm, we arrive at A, then we cannot proceed by adding to A any element from $B - A$.

We construct a weight function designed to assure that when we apply the greedy algorithm we have the following results: 1) the greedy algorithm first picks all elements in $A \cap B$; 2) next, the algorithm picks all elements in $A - B$ (these two steps effectively pick all elements in A); 3) the algorithm would prefer to choose elements from $B - A$ if possible; 4) the algorithm picks other elements.

In constructing such a weight function it is convenient to make the following definitions:

$$D_A = |A - B| \quad D_B = |B - A|.$$

In addition, recalls that given two sets we may divide the elements of the sets into elements that are shared between the sets and elements that are not shared so that

$$|A| = |A \cap B| + |A - B| \quad |B| = |A \cap B| + |B - A|.$$

Since $|A| < |B|$, we know that $|A - B| < |B - A|$, or $D_A < D_B$.

Now, construct the weight function as:

$$w(e) = \begin{cases} 2 & e \in A \cap B \\ 1/D_A & e \in A - B \\ (1 + \varepsilon)/D_B & e \in B - A \\ 0 & otherwise \end{cases}$$

This weight function accomplishes exactly what we set out to do. When we start the greedy algorithm, we will first choose all elements in $A \cap B$, then the elements in $A - B$. The algorithm will want to choose from $B - A$ next, but because no member of $B - A$ can be added, the algorithm is forced to choose the remainder of elements elsewhere.

Now, compare the weight of the maximal element obtained from the greedy algorithm, G, with the weight of the maximal element $\bar{B} \supseteq B$. The weight of the solution from the greedy algorithm is

$$\Omega(G) = 2|A \cap B| + \frac{|A - B|}{D_A} = 2|A \cap B| + 1$$

The greedy algorithm first picks all elements of $A \cap B$, and each of these has a weight of 2. Next, it picks all elements of $A - B$, and these have weight $1/D_A$. Since it cannot pick any of the elements in $B - A$, the algorithm proceeds to other elements, which have a weight of 0.

The weight of $\bar{B} \supseteq B$ is

$$\Omega(\bar{B}) \geq \Omega(B) = 2|A \cap B| + \frac{(1+\varepsilon)}{D_B}|A - B| = 2|A \cap B| + 1 + \varepsilon$$

In this case, the weight of $\bar{B}$ must be at least as large as the weight of B because all of the weight components have non-negative value. The weight of B is the sum of the weights of the elements in B, leading to the above result.

However, comparing the results we see

$$\Omega(\bar{B}) - \Omega(G) > \Omega(B) - \Omega(G) = 1 + \varepsilon - 1 = \varepsilon > 0$$

In this case, there is a maximal element that has a greater weight than the solution from the greedy algorithm. This contradicts the assertion **GA-3** that the greedy algorithm produces a maximal element with maximum weight.

Thus, it must be the case that **GA-3** implies **SM-3**. Therefore, the system (S, T) is a matroid. ∎

6.1.5 COROLLARY

The greedy algorithm produces a maximal element of maximum weight for the system (S, T) if and only if (S, T) is a matroid.

Proof. This is simply the combinations of the previous two theorems. Theorem 6.1.3 proved that if (S, T) is a matroid, the greedy algorithm produces a maximal element of maximum weight. Moreover, Theorem 6.1.4 showed that if the greedy algorithm works for a system (S, T), then it produces a maximal element of maximum weight. ∎

6.1.6 COROLLARY

If (S, T) is a matroid, then sequence of sets $X_1, X_2, \ldots, X_n$ obtained from the greedy algorithm is the maximal weight subset for every value of $1 \leq k \leq n$.

Proof. The proof is largely unchanged from Theorem 6.1.3. There we showed that the solution from the greedy algorithm is the maximum value for any sequence of length r. There is nothing in the proof restricting r to be the rank of the matroid, so this must be true for every value of r leading to a maximal element.

Thus, if (S, T) is a matroid, the greedy algorithm not only produces a maximal element of maximum weight, each element in the sequence is maximum weight among element of the same dimension. ∎

6.1.7 THEOREM

Let M be a matroid over the set S with independent set T. Let $X, Y \subseteq S$ where $|X| = |Y| = k$. Order the elements of X and Y by decreasing weight $w(X_i)$ and $w(Y_i)$. The greedy algorithm will operate successfully if the weight function $\Omega(X)$ obeys

WF-1   If $w(X_i) \geq w(Y_i)$ for every i, then $\Omega(X) \geq \Omega(Y)$.

Proof. Examining theorem 6.1.3, the proof requires that if an ordered set of element obeys $w(X_i) \geq w(Y_i)$, then $\Omega(X) \geq \Omega(Y)$. In order for this proof to function, we do not need the weight function $\Omega(X)$ to be the sum of the $w(X_i)$'s, we just need the weight function to satisfy $w(X_i) \geq w(Y_i) \rightarrow \Omega(X) \geq \Omega(Y)$.

Alternatively, theorem 6.1.4 has the same requirement. So as long as $w(X_i) \geq w(Y_i) \rightarrow \Omega(X) \geq \Omega(Y)$ is true, the proof of 6.1.4 will operate.

In each case, the only requirement for the proofs to operate is that $w(X_i) \geq w(Y_i) \rightarrow \Omega(X) \geq \Omega(Y)$. We can extend the greedy algorithm to include a broader set of weight functions. For example, if the weight function on elements satisfies $0 < w(x) < \infty$, then the product weight function satisfies $w(X_i) \geq w(Y_i) \rightarrow \Omega(X) \geq \Omega(Y)$:

$$\Omega(X) = \prod_{x \in X} w(x). \quad \blacksquare$$

6.1.8   DEFINITION: KRUSKAL'S ALGORITHM

> Let G be a graph and E the edge set of G. Assign a $w(E)$ to each edge.
>
> We desire to find a spanning tree of G with minimal weight. Proceed as follows:
>
> 1.   Choose an edge with minimal weight as X_n
> 2.   Find the set N of all edges $e \in E - X_n$ such that $e \cup X_n$ does not contain a cycle in G.
>   a.   If $N \neq \emptyset$, choose $n \in N$ with $w(n)$ maximum. Set $X_{n+1} = e \cup X_n$. Repeat step 2.
>   b.   If $N = \emptyset$, then X_n is the desired spanning tree.

Kruskal's algorithm is a particular instance of the greedy algorithm. The choice of e in step 2 to avoid creating a cycle in G is the same as choosing $s \cup X_n \in T$ in step 2 of the greedy algorithm.

In the case of graphs, the cycles of a graph form the dependent sets of the matroid. Thus, all independent sets of the matroid do not contain a cycle. By avoiding the creation of a cycle, Kruskal's algorithm restricts the edge choices to the independent set, just as in the greedy algorithm.

6.1.9   EXAMPLE

Find the minimal weight spanning tree for the graph in Figure 29.

We apply Kruskal's algorithm. The edges of the graph are labeled with their weight. We first choose the edge with weight 1. We can add any edge of the graph to this without creating a cycle, so we choose the edge with weight 2. Again, we can add any edge to the graph, so we choose the edge with weight 3.

Once we arrive at the edge set $X_3 = \{1,2,3\}$, we are not able to add an arbitrary edge without creating a cycle. If we add edge 4 to this set, we have a cycle. However, we can add any other edge. We choose the edge with minimum weight, edge 5. At this point we are unable to add any more edges without creating a cycle. The desired solution is $\{1,2,3,5\}$.

6.1.10 DEFINITION: MATRIC MATROID ALGORITHM

Let K be a $n \times m$ matrix with column headers C. Assign a weight $w(C)$ to each column.

We desire to find a maximal independent set X of column vectors whose total weight is a maximum:

1. Order the columns from left to right by weight so $w(v_1) \geq w(v_2) \geq \cdots \geq w(v_n)$.
2. Choose the highest weight non-zero column e not in X_n.
   a. Set $X_{n+1} = e \cup X_n$. Use e to eliminate nonzero entries in the columns to the right. Repeat step 2.
   b. If there is no e satisfying 2, then X_n is the desired spanning tree.

The matric matroid algorithm is another particular instance of the greedy algorithm. The manipulation of the columns to the right in step 2a is done to avoid creating a dependent cycle in X. This is the same as choosing $s \cup X_n \in T$ in step 2 of the greedy algorithm.

6.1.11 EXAMPLE

Find the maximum weight set of independent columns for the matrix

$$\begin{bmatrix} 1 & 1 & 0 & 1 & 0 \\ 0 & 0 & 0 & 1 & 1 \\ 0 & 1 & 1 & 1 & 0 \end{bmatrix}$$

Weight: 5   4   3   2   1

The columns are already ordered by weight. We begin by initially choosing $X_1 = \{v_1\}$. Next, we eliminate column entries by subtracting v_1 from the other columns. The new matrix is

$$\begin{bmatrix} 1 & 0 & 0 & 0 & 0 \\ 0 & 0 & 0 & 1 & 1 \\ 0 & 1 & 1 & 1 & 0 \end{bmatrix}$$

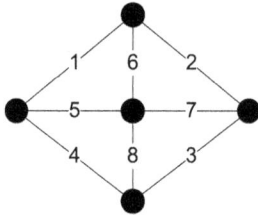

Figure 29: Weighted graph.   Figure 30: Minimum spanning tree.

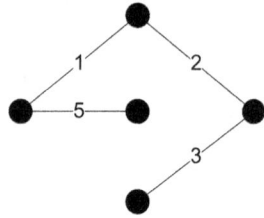

Next we choose the highest weight non-zero column, which is v_2. We have

$$X_2 = X_1 \cup v_2 = \{v_1, v_2\}.$$

Next we use v_2 to eliminate entries to the right of v_2. The new matrix is

$$\begin{bmatrix} 1 & 0 & 0 & 0 & 0 \\ 0 & 0 & 0 & 1 & 1 \\ 0 & 1 & 0 & 0 & 0 \end{bmatrix}$$

The highest weight non-zero column is v_4. We add this to the set and eliminate which gives

$$X_3 = X_2 \cup v_4 = \{v_1, v_2, v_4\}$$

and

$$\begin{bmatrix} 1 & 0 & 0 & 0 & 0 \\ 0 & 0 & 0 & 1 & 0 \\ 0 & 1 & 0 & 0 & 0 \end{bmatrix}.$$

There are no non-zero columns left, so we are finished. The final set is $X = \{v_1, v_2, v_4\}$. This is the maximal independent set with maximal weight.

6.1.12 DEFINITION: CIRCUIT ALGORITHM

> Let $M = (S, C)$ be a matroid over a set S with circuits (cycles) C. Let $w(s_1) \geq w(s_2) \geq \cdots \geq w(s_n)$ be weights assigned to the elements $s_i \in S$. We desire to find a maximal independent set X with maximum weight.
>
> 1. Construct the set $A_0 = \emptyset, B_0 = S$.
> 2. Choose any circuit $c \in C$ such that $c \subseteq B$. If no such element exists, terminate.
> 3. Find the highest weighted element $e \in c$.
>   a. Construct $B_{n+1} = B_n - e$
>   b. Construct $A_{n+1} = A_n \cup e$
>
> The set A is the maximal independent set with maximum weight. The set B is the maximal independent set with minimum weight.

In some situations, especially when dealing with graphs, it is easier to identify the circuits of the matroid instead of the independent sets. In this case, the circuit algorithm may be used to identify the maximal independent set with maximum or minimum weight.

6.1.13 EXAMPLE

The cycles for the graph in Figure 29 are:

{1,2,3,4} {1,2,3,5,8} {1,2,4,7,8} {1,2,5,7} {1,3,4,6,7} {1,4,6,8}

{1,5,6} {2,3,4,5,6} {2,3,6,8} {2,6,7} {3,4,5,7} {3,8,7} {4,5,8}

Find the minimal weight spanning tree.

Start with $A_0 = \emptyset$, $B_0 = \{1,2,3,4,5,6,7,8\}$. We can begin with any cycle, pick $\{1,2,3,4\}$. The highest weight element is 4. We move 4 from B and put it in A so

$$A_1 = A_0 \cup \{4\} = \{4\}$$

$$B_1 = B_0 - \{4\} = \{1,2,3,5,6,7,8\}.$$

Since $\{4\}$ is no longer in B, we need to choose a cycle that does not contain $\{4\}$. The first such cycle on the list is $\{1,2,3,5,8\}$. The highest weight element is 8. Constructing A and B for the next iteration,

$$A_2 = A_1 \cup \{8\} = \{4,8\}$$

$$B_2 = B_1 - \{8\} = \{1,2,3,5,6,7\}.$$

Next we need to choose a cycle containing only elements from the set $B_2 = \{1,2,3,5,6,7\}$. The first such set is $\{1,2,5,7\}$. Repeating the previous steps,

$$A_3 = A_2 \cup \{7\} = \{4,7,8\}$$

$$B_3 = B_2 - \{7\} = \{1,2,3,5,6\}.$$

The next cycle on this list where $c \subseteq B_3$ is $\{1,5,6\}$. Again, repeating the augmentation steps,

$$A_4 = A_3 \cup \{6\} = \{4,6,7,8\}$$

$$B_4 = B_3 - \{6\} = \{1,2,3,5\}.$$

None of the remaining cycles satisfy $c \subseteq B_4$. This ends the algorithm. A_4 is the maximal independent set with maximum weight, while B_4 is the maximal independent set with minimum weight. Moreover, we can confirm that we have arrived at the same set as in Example 6.1.9.

6.1.14 PROBLEMS

1. Let $S = \{a, b, c, d\}$ and $T = \{\emptyset, \{a\}, \{b\}, \{c\}, \{a, b\}, \{a, c\}, \{b, c\}\}$ with weights $w(\emptyset) = 0, w(\{a\}) = 1, w(\{b\}) = 2, w(\{c\}) = 3, w(\{d\}) = 4$. Use the greedy algorithm to find the maximal member of T. Is the maximal member the global optimum?

2. Let $S = \{a, b, c, d\}$ and $T = \{\emptyset, \{a\}, \{b\}, \{c\}, \{d\}, \{a, b\}, \{a, c\}, \{b, c\}\}$ with weights $w(\emptyset) = 0, w(\{a\}) = 1, w(\{b\}) = 2, w(\{c\}) = 3, w(\{d\}) = 4$. Use the greedy algorithm to find the maximal member of T. Is the maximal member the global optimum?

6.2   Partitions

Some problems can be analyzed with a maximum cardinality partition of some set of matroids. This section examines matroid partitions and presents an algorithm designed to find a maximum cardinality partition of a set of matroids.

6.2.1   DEFINITION: MATROID PARTITIONING ALGORITHM

Let $M_1, M_2, \ldots M_n$ be matroids (S, T_n) over the same ground set S. Let $r_l(X)$ be the rank function for each matroid, $l = 1, 2, \ldots n$. Similarly, let $cl_l(X)$ be the closure operator for each matroid, $l = 1, 2, \ldots n$. We desire to find a partition of S into n sets I_l such that $I_l \in T_l$.

1. Construct the sets $I_l = \emptyset$ for $l = 1, 2, \ldots n$
2. Construct a set $U = S$
3. Set $S_0 = S$ and set $j = 1$
4. Find the least l such that $|I_l \cap S_{j-1}| < r_l(S_{j-1})$
   a. If no such l exists, there is no partition; halt
5. Set $S_j = S_{j-1} \cap cl_l(I_l \cap S_{j-1})$
6. Set $h(j) = l$
7. If $U \subseteq S_j$, set $j = j + 1$. Goto Step 3
   a. Otherwise, let $e \in U - S_j$ be any element of $U - S_j$
8. Set $U = U - e$
9. Set $I_{h(j)} = I_{h(j)} \cup e$
10. If $I_{h(j)}$ is independent in $M_{h(j)}$, goto Step 12
11. Find the unique dependent set $C \subseteq I_{h(j)}$. Let $e' \in C - S_j$ be any element of $C - S_j$.
    a. Set $I_{h(j)} = I_{h(j)} - e'$
    b. Set $e = e'$
    c. Set $j = j - 1$
    d. Goto Step 9
12. If $U \neq \emptyset$ then goto Step 3.
    a. Otherwise, all elements are assigned; halt

The partitioning algorithm augments each partition as much as possible, then moves on to the next partition. If a partition is augmented to become a dependent set, the algorithm removes an element (making the set independent again), and the removed element is put into a previous partition.

6.2.2 EXAMPLE

We illustrate the partitioning algorithm with a specific example. This example may be skipped if the reader is not interested in the specific operation of the partitioning algorithm. However, this example may be useful in checking computer programs implementing the algorithm.

The algorithm is designed to repeat until each element is placed into a partition. Some of the data is transitory and is only valid within a specific loop, while other data persists between iterations.

At each step we place the data values into two tables. The table on the left is persistent data, while the table on the right is transitory data.

Find a partition of the matroids M_1, M_2 defined as

$$M_1 = \begin{cases} S = \{a, b, c, d\} \\ B = \{\{a, b, c\}\} \end{cases} \quad M_2 = \begin{cases} S = \{a, b, c, d\} \\ B = \{\{c, d\}\} \end{cases}$$

Steps 1-2:

These are initialization steps for the persistent data. We initialize each of the I_l's to the empty set, and set $U = S$. We do not come back to this step again.

Persistent	
U	$\{a, b, c, d\}$
I_1	$\emptyset$
I_2	$\emptyset$
$h(1)$	

Step 3:

This step initializes the transitorily data for first loop. We start with $S_0 = S$ and set $j = 1$.

Persistent	
U	$\{a, b, c, d\}$
l_1	$\emptyset$
l_2	$\emptyset$
$h(1)$	

Transitory	
S_0	$\{a, b, c, d\}$
j	1

Step 4:

We need to find the least l such that $|l_l \cap S_0| < r_l(S_0)$. Start with l_1.

$$|l_1 \cap S_0| = |\emptyset \cap S_0| = |\emptyset| = 0$$
$$r_1(S_0) = r_1(\{a, b, c, d\}) = 3$$

We see that $l = 1$ satisfies the inequality. Set this in the transitory data.

Persistent	
U	$\{a, b, c, d\}$
l_1	$\emptyset$
l_2	$\emptyset$
$h(1)$	

Transitory	
S_0	$\{a, b, c, d\}$
j	1
l	1

Step 5:

Set $S_1 = S_0 \cap cl_1(l_1 \cap S_0)$. The closure is

$$cl_1(l_1 \cap S_0) = cl_1(\emptyset \cap S_0) = cl_1(\emptyset) = \{d\}$$

Substituting the closure into the expression for S_1,

$$S_1 = S_0 \cap \{d\} = \{d\}$$

Persistent	
U	$\{a, b, c, d\}$
l_1	$\emptyset$
l_2	$\emptyset$
$h(1)$	

Transitory	
S_0	$\{a, b, c, d\}$
j	1
l	1
S_1	$\{d\}$

Step 6:

Update the persistent data by setting $h(1) = 1$.

Persistent	
U	$\{a, b, c, d\}$
l_1	$\emptyset$
l_2	$\emptyset$
$h(1)$	1

Transitory	
S_0	$\{a, b, c, d\}$
j	1
l	1
S_1	$\{d\}$

Step 7:

Currently, $U = \{a, b, c, d\}$ while $S_1 = \{d\}$. Thus, $U \nsubseteq S_1$.

In this case we execute step 7a. First, construct $U - S_1 = \{a, b, c\}$. Next, we choose any element e from $\{a, b, c\}$. Choose $e = a$. Update the transitory data with this choice for e.

Persistent	
U	$\{a, b, c, d\}$
I_1	$\emptyset$
I_2	$\emptyset$
$h(1)$	1

Transitory	
S_0	$\{a, b, c, d\}$
j	1
l	1
S_1	$\{d\}$
e	a

Step 8:

We update the value of U by setting $U = U - e = \{b, c, d\}$.

Persistent	
U	$\{b, c, d\}$
I_1	$\emptyset$
I_2	$\emptyset$
$h(1)$	1

Transitory	
S_0	$\{a, b, c, d\}$
j	1
l	1
S_1	$\{d\}$
e	a

Step 9:

Now we update the value of I_1 by setting $I_1 = I_1 \cup e = \{a\}$.

Persistent	
U	$\{b, c, d\}$
I_1	$\{a\}$
I_2	$\emptyset$
$h(1)$	1

Transitory	
S_0	$\{a, b, c, d\}$
j	1
l	1
S_1	$\{d\}$
e	a

Step 10:

I_1 is an independent set in M_1, so we move to step 12.

Step 12:

$U = \{b, c, d\} \neq \emptyset$, so we return to step 3.

Step 3:

We reinitialize the loop by setting $S_0 = S$ and set $j = 1$. These were not updated during the past iteration, so there is in fact no change to the transitory data.

Persistent	
U	$\{b,c,d\}$
I_1	$\{a\}$
I_2	$\emptyset$
$h(1)$	1

Transitory	
S_0	$\{a,b,c,d\}$
j	1
l	1
S_1	$\{d\}$
e	a

Step 4:

Again, we seek the least l such that $|I_l \cap S_0| < r_l(S_0)$. This is satisfied by I_1.

$$|I_1 \cap S_0| = |\{a\} \cap S_0| = |\{a\}| = 1$$
$$r_1(S_0) = r_1(\{a,b,c,d\}) = 3$$

Setting $l = 1$ in the transitory data,

Persistent	
U	$\{b,c,d\}$
I_1	$\{a\}$
I_2	$\emptyset$
$h(1)$	1

Transitory	
S_0	$\{a,b,c,d\}$
j	1
l	1
S_1	$\{d\}$
e	a

Step 5:

Set $S_1 = S_0 \cap cl_1(I_1 \cap S_0)$. The closure is

$$cl_1(I_1 \cap S_0) = cl_1(\{a\} \cap S_0) = cl_1(\{a\}) = \{a,d\}$$

Substituting the closure into the expression for S_1,

$$S_1 = S_0 \cap \{a,d\} = \{a,d\}$$

Persistent	
U	$\{b,c,d\}$
I_1	$\{a\}$
I_2	$\emptyset$
$h(1)$	1

Transitory	
S_0	$\{a,b,c,d\}$
j	1
l	1
S_1	$\{a,d\}$
e	a

Step 6:

$h(1) = 1$ already, so no update is required.

Steps 7-9:

U is not a subset of S_1, so we need to choose $e \in U - S_1 = \{b,c\}$. Choose $e = b$. With this choice we update $U = U - e = \{c,d\}$ and $I_1 = I_1 \cup e = \{a,b\}$.

Persistent	
U	$\{c,d\}$
I_1	$\{a,b\}$
I_2	$\emptyset$
$h(1)$	1

Transitory	
S_0	$\{a,b,c,d\}$
j	1
l	1
S_1	$\{a,d\}$
e	b

Step 10:

I_1 is an independent set in M_1, so we move to step 12.

Step 12:

$U = \{c,d\} \neq \emptyset$, so we return to step 3.

Steps 3-12:

The next loop will proceed similarly to the previous two. We will move through each of the past steps with similar results. At the end of step 12 we have the following data:

Persistent	
U	$\{d\}$
I_1	$\{a,b,c\}$
I_2	$\emptyset$
$h(1)$	1

Transitory	
S_0	$\{a,b,c,d\}$
j	1
l	1
S_1	$\{a,b,d\}$
e	c

Again, $U \neq \emptyset$, so we return to step 3.

Steps 3-4:

Step 3 simply reinitializes the transitory variables to start the next loop. In this iteration, step 4 is different than the past loops because this time $l = 1$ does not satisfy $|I_l \cap S_0| < r_l(S_0)$. Consequently, we use $l = 2$.

Updating the transitory data,

Persistent	
U	$\{d\}$
I_1	$\{a,b,c\}$
I_2	$\emptyset$
$h(1)$	1

Transitory	
S_0	$\{a,b,c,d\}$
j	1
l	2
S_1	$\{a,b,d\}$
e	c

Steps 5-6:

$$S_1 = S_0 \cap cl_2(I_2 \cap S_0) = S_0 \cap cl_2(\emptyset) = S_0 \cap \{a,b\} = \{a,b\}$$

$$h(1) = 2$$

Persistent	
U	$\{d\}$
I_1	$\{a, b, c\}$
I_2	$\emptyset$
$h(1)$	2

Transitory	
S_0	$\{a, b, c, d\}$
j	1
l	2
S_1	$\{a, b\}$
e	c

Steps 7-9:

U is not a subset of S_1, so we need to choose $e \in U - S_1 = \{d\}$.
The only choice is $e = d$. Update $U = U - e = \emptyset$ and $I_2 = I_2 \cup e = \{d\}$.

Persistent	
U	$\emptyset$
I_1	$\{a, b, c\}$
I_2	$\{d\}$
$h(1)$	2

Transitory	
S_0	$\{a, b, c, d\}$
j	1
l	2
S_1	$\{a, b\}$
e	d

Step 10:

I_2 is an independent set in M_2, so we move to step 12.

Step 12:

Finally, we have $U = \emptyset$. All elements have been assigned and we have found a maximum cardinality partition of S:

$$I_1 = \{a, b, c\}$$
$$I_2 = \{d\}$$

6.3   Matroid Intersections

Many problems can be reduced to an intersection of two matroids. This section examines matroid intersections and algorithms designed to find optimal independent sets of intersection matroids.

6.3.1 DEFINITION: AUGMENTING ALGORITHM

Let M_1 be the matroid (S, T_1) and M_2 the matroid (S, T_2). Let I be any intersection of M_1 and M_2. We desire to find a set I' where $|I'| > |I|$. We construct an augmenting sequence for I using the following steps:

13. Construct the sets $I_0 = I, J_0 = S$.
14. Select an element $t \in J_n$ such that $I_n \cup t \in T_1$. If no such element exists, move to step 3.
15. If $I_n \cup t \in T_2$, then I_n is an augmented sequence for I, and terminate the algorithm. Otherwise, there is a unique cycle in T_2 from the elements $I_n \cup t$. Find an element $u \in (J_n - t)$ such that $I_n \cup t - u \in T_2$. Set $I_{n+1} = I_n \cup t$ and $J_{n+1} = J_n - t$.
16. If $J_{n+1} = \emptyset$, no augmenting sequence exists, terminate the algorithm. Otherwise, return to step 2.

The augmentation algorithm attempts to find an independent set of both M_1 and M_2 that is larger than the initial set I. If the sequence terminates without finding such a sequence, then no such sequence exists.

6.3.2 DEFINITION: SPAN

Let M_1 be the matroid (S, T). Let $P \subseteq S$ be a subset of S. The span of P, denoted $sp(P)$, is the maximal superset of $R \supseteq P$ such that the rank of R equals the rank of P.

In a graph, the span is the set of all single edges that can be added to P and create a circuit. Let C be the set of edges such that for every $c \in C$, $P \cup c \notin T$. Then $sp(P) = P \cup C$.

6.3.3 DEFINITION: CARDINALITY INTERSECTION ALGORITHM

Let M_1 be the matroid (S, T_1) and M_2 the matroid (S, T_2). Let I be any intersection of M_1 and M_2. We desire to find a set I' where $|I'| > |I|$. We construct an augmenting sequence for I using the following steps:

1. For every $e_i \in S - I$, find the minimum circuits C_i^1 and C_i^2 created by the set $I \cup e_i$ in M_1 and M_2 respectively.
2. Label each element $e_i \in S - sp_1(I)$ as $\emptyset^+$.
3. Find an unchecked element $e_k \in S$. Mark e_k as checked. If no labels are unchecked, proceed to step 7.
4. If the label is '+' and $I \cup e_k \in T_2$, proceed to step 6. Otherwise label every unlabeled element in C_k^2 with 'k^-'. Return to step 3.
5. If the label is '−', label every unlabeled element e_j with the label 'k^-' if $e_k \in C_j^1$. Return to step 3.
6. We have discovered an augmenting sequence. We have labeled each node with as $k^\pm$, i.e. $1^\pm, 2^\pm$, etc. We can use this to trace the final entry back to an origin labeled $\emptyset^+$.
   a. Construct this sequence.
   b. Augment I by adding every element in the sequence marked with a '+' label, and removing those marked with a '−' label. This is the augmented set I'.
   c. Clear all labels and return to step 1.
7. No augmenting sequence exists. Terminate the algorithm.

The cardinality intersection algorithm tests an independent set to determine if it has maximum cardinality. If a higher cardinality set is found, the algorithm is repeated. If no such set is found, maximal cardinality is reached by the original set.

The algorithm uses a series of labels applied to the common ground set of the matroids. There are two main labels: + and −. Once an element is labeled it must be checked. The algorithm either iterates once a higher cardinality set is found or terminates once all labels are checked.

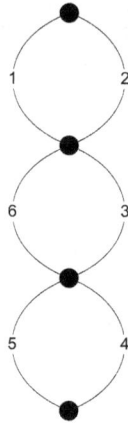

Figure 31: M_1 for matroid cardinality intersection.

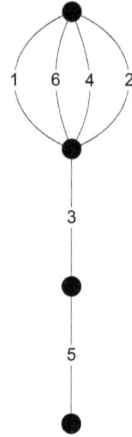

Figure 32: M_2 for matroid cardinality intersection.

6.3.4 EXAMPLE

Let M_1 and M_2 be the matroids arising from the graphs in Figure **31** and Figure **32** respectively. Find the maximum cardinality intersection of these matroids starting from $I = \{3,4\}$.

The initial set I is a maximal intersection in the sense that no element can be added to I without creating a cycle in either M_1 or M_2. However, we can readily verify that there are three element intersections between the matroids. We apply the cardinality intersection algorithm and expect to discover one such intersection.

Step 1 Computations:

$$I = \{3,4\} \quad sp_1(I) = \{3,4,5,6\} \quad S - sp_1(I) = \{1,2\}$$

$$C_5^1 = \{4,5\} \quad \begin{matrix} C_1^2 = \{1,4\} \\ C_2^2 = \{2,5\} \\ C_6^1 = \{3,6\} \quad C_6^2 = \{4,6\} \end{matrix}$$

Step 2 Labeling:

We need to label the elements 1 and 2 with the label $\emptyset^+$.

1	2	3	4	5	6
$\emptyset^+$	$\emptyset^+$				

Step 3:

Start with element $k = 1$. This element hasn't been checked, so we check it and mark it with an overbar ($\bar{1}$) to indicate that it has been checked.

$\bar{1}$	2	3	4	5	6
$\emptyset^+$	$\emptyset^+$				

Step 4:

Element 1 is labeled as a '+', so we check if $I \cup e_1 \in T_2$. However, $I \cup e_1 \notin T_2$, so we examine C_1^2 for unlabeled elements. C_1^2 has two elements: element 4 is unlabeled, while element 1 is labeled. We label element 4 with the label 1^- ($k = 1$), then return to step 3.

$\bar{1}$	2	3	4	5	6
$\emptyset^+$	$\emptyset^+$				
			1^-		

Step 3:

Next choose element $k = 2$. This element hasn't been checked, so we check it and mark it with an overbar to indicate that it has been checked.

$\bar{1}$	$\bar{2}$	3	4	5	6
$\emptyset^+$	$\emptyset^+$				
			1^-		

Step 4:

Element 2 is labeled as a '+', so we check if $I \cup e_2 \in T_2$. However, $I \cup e_2 \notin T_2$, so we examine C_2^2 for unlabeled elements. C_2^2 has two elements, 4 and 1, but both are labeled. We return to step 3.

Step 3:

Next choose element $k = 4$. We do not choose $k = 3$ because we can only check labeled elements. Again, we mark the element as checked.

$\bar{1}$	$\bar{2}$	3	$\bar{4}$	5	6
$\emptyset^+$	$\emptyset^+$				
			1^-		

Step 4:

Element 4 is labeled as a $-$, so step 4 does not apply. We proceed to step 5.

Step 5:

We need to label every e_j as 4^+ if $e_4 \in C_j^1$. C_5^1 is the only one that satisfies this, so we mark element 5, then return to step 3.

$\bar{1}$	$\bar{2}$	3	$\bar{4}$	5	6
$\emptyset^+$	$\emptyset^+$				
			1^-		
				4^+	

Step 3:

Next choose element $k = 5$ and mark the element as checked.

$\bar{1}$	$\bar{2}$	3	$\bar{4}$	$\bar{5}$	6
$\emptyset^+$	$\emptyset^+$				
			1^-		
				4^+	

Step 4:

In this case we find $I \cup e_5 \in T_2$, so we proceed to step 6.

Step 6 Augmentation:

This is an augmented sequence. From the labeling, we can trace the sequence back to its origin: $5 \rightarrow 4 \rightarrow 1$.

The augmented set is

$I' = I + \{1\} - \{4\} + \{5\} = \{1,3,5\}.$

To proceed, we clear all labels and restart with step 1.

Step 1 Computations:

$$I = \{1,3,5\} \quad sp_1(I) = \{1,2,3,4,5,6\} \quad S - sp_1(I) = \emptyset$$

Step 2 Labeling:

This time the test set spans the matroid. There are no elements to label as $\emptyset^+$.

Step 3:

There are no unchecked labeled elements, so we proceed to step 7.

Step 7:

The set $\{1,3,5\}$ is a maximum cardinality intersection of the matroids. This can be verified by examining Figure 31 and Figure 32. We see in each figure that this set of edges does not form a cycle in either graph. Furthermore, we cannot add another edge to this set without creating a cycle in at least one of the graphs (both in this example).

Chapter 7: Enumeration

This chapter examines the problem of enumerating all matroids for a given number of elements. In general, matroids have many different presentations in the sense that the set of bases or independent sets are not the same but have the same structure.

For instance, examine the matroid

$$S = \{a, b\}$$

$$B = \{\{a\}\}$$

This is the same object as the matroid

$$S = \{x, y\}$$

$$B = \{\{y\}\}$$

These different presentations are not considered different matroids because they have the same underlying mathematical structure. We proceed to create an algorithm capable of identifying the distinct matroids on a given number of elements.

7.1   Isomorphism and Presentations

Two matroids may have the same underlying structure even though their set of bases may appear different. In this case the matroids are called isomorphic, and the different bases result from permutations of the ground set.

In enumerating matroids, it is customary to only count matroids that have a different structure rather than just a different presentation of the ground set.

7.1.1 DEFINITION: ISOMORPHIC MATROIDS

> Let $M_1 = (S, B_1)$ and $M_2 = (S, B_2)$ be matroids on the ground set S with bases B_1 and B_2 respectively. M_1 and M_2 are called isomorphic if there exists some permutation P on S such that $PB_1 = B_2$.

7.1.2 EXAMPLE

Let $M_1 = (S, B_1)$ where $S = \{a, b, c, d, e, f\}$ and $B_1 = \{\{a, b\}\}$. Let $M_2 = (S, B_2)$ where $S = \{a, b, c, d, e, f\}$ and $B_2 = \{\{e, f\}\}$.

Examine the action of the permutation

$$P = \begin{pmatrix} a & b & c & d & e & f \\ e & f & c & d & a & b \end{pmatrix}$$

Effectively, this permutation exchanges $a \leftrightarrow e$ and $b \leftrightarrow f$. If we make this exchange everywhere in $B_1 = \{\{a, b\}\}$ we get $PB_1 = \{\{e, f\}\}$ which is B_2.

M_1 and M_2 are isomorphic matroids because the permutation $P = \begin{pmatrix} a & b & c & d & e & f \\ e & f & c & d & a & b \end{pmatrix}$ maps $B_1 \rightarrow B_2$.

7.1.3 DEFINITION: PRESENTATIONS

> Let $M = (S, B)$ be a matroid on the ground set S with bases B. A presentation of M is found by applying a permutation P to S.

There are $|S|!$ permutations that may be applied to S. Each permutation may lead to a unique presentation of M, but often many different permutations arrive at the same presentation.

7.1.4 EXAMPLE

Find all presentations of the matroid $M = (S, B)$ with ground set $S = \{a, b\}$ and bases $B = \{\{a\}\}$.

There are two permutations on a set of two elements:

$$P_1 = \begin{pmatrix} a & b \\ a & b \end{pmatrix} \quad P_2 = \begin{pmatrix} a & b \\ b & a \end{pmatrix}$$

This first permutation does make any changes, while the second exchanges $a \leftrightarrow b$. Applying each to the bases:

$$P_1 B = \{\{a\}\} \quad P_2 B = \{\{b\}\}$$

We have two different presentations for this matroid, and each permutation leads to a unique presentation.

7.1.5   EXAMPLE

Find all presentations of the matroid $M = (S, B)$ with ground set $S = \{a, b, c\}$ and bases $B = \{\{a, b\}\}$.

There are six permutations on a set of three elements:

$$P_1 = \begin{pmatrix} a & b \\ a & b \end{pmatrix} \quad P_1 = \begin{pmatrix} a & b \\ b & a \end{pmatrix}$$

$$P_1 = \begin{pmatrix} a & b & c \\ a & b & c \end{pmatrix} \quad P_2 = \begin{pmatrix} a & b & c \\ a & c & b \end{pmatrix} \quad P_3 = \begin{pmatrix} a & b & c \\ b & a & c \end{pmatrix}$$

$$P_4 = \begin{pmatrix} a & b & c \\ b & c & a \end{pmatrix} \quad P_5 = \begin{pmatrix} a & b & c \\ c & a & b \end{pmatrix} \quad P_6 = \begin{pmatrix} a & b & c \\ c & b & a \end{pmatrix}$$

Applying each to the bases:

$$P_1 B = \{\{a, b\}\} \quad P_2 B = \{\{a, c\}\} \quad P_3 B = \{\{b, a\}\}$$

$$P_4 B = \{\{b, c\}\} \quad P_5 B = \{\{c, a\}\} \quad P_6 B = \{\{c, b\}\}$$

The order of the elements inside of the bases is immaterial, so $P_1 B = P_3 B$. Similarly, $P_2 B = P_5 B$ and $P_4 B = P_6 B$. With this, we see there are three different presentations of this matroid with bases: $B = \{\{a, b\}\}$, $B = \{\{a, c\}\}$, and $B = \{\{b, c\}\}$.

7.2   Augmentation and Extension

Augmentation and extension are operations that increase the number of elements of the ground set of a matroid. These operations produce new matroids on a larger ground set from a smaller matroid.

7.2.1 DEFINITION: AUGMENTED MATROID

> Let M be a matroid with ground set S and set of bases B. M is augmented if there is some $e \in S$ such that $e \in b$ for every $b \in B$.

Augmented matroids are matroids where some element e is present in every basis.

7.2.2 DEFINITION: EXTENDED MATROID

> Let M be a matroid with ground set S and set of bases B. M is extended if there is some $e \in S$ such that $e \notin b$ for every $b \in B$.

Extended matroids are matroids where some element e is not present in every basis.

7.2.3 EXAMPLE

The matroid with ground set $S = \{a, b, c, d, e\}$ and set of bases

$$B = \begin{cases} \{a, c, e\} \\ \{a, d, e\} \\ \{b, c, e\} \\ \{b, d, e\} \\ \{c, d, e\} \end{cases}$$

is augmented in e because e appears in every basis of B.

7.2.4 EXAMPLE

The matroid with ground set $S = \{a, b, c, d, e\}$ and set of bases

$$B = \begin{cases} \{a, c\} \\ \{a, d\} \\ \{b, c\} \\ \{b, d\} \\ \{c, d\} \end{cases}$$

is extended in e because e does not appear in any basis of B.

7.2.5 DEFINITION: AUGMENTATION OPERATOR

Let $\mathfrak{M}_n$ be the set of matroids on n elements with ground set $S = \{e_1, e_2, \dots, e_n\}$. Let $\mathcal{A}: \mathfrak{M}_n \to \mathfrak{M}_{n+1}$ be an operator acting on an element $M_n \in \mathfrak{M}_n$ with bases B_n such that $\mathcal{A}M_n = M_{n+1}$ where $M_{n+1} \in \mathfrak{M}_{n+1}$ and M_{n+1} is the matroid where the bases B_{n+1} satisfy

AO-1 $|B_{n+1}| = |B_n|$

AO-2 For every $b_{n+1} \in B_{n+1}$ there exists some $b_n \in B_n$ such that $b_{n+1} = b_n \cup e_{n+1}$

The augmentation operator creates a new matroid by adding an additional element e_{n+1} to every basis of the original matroid.

7.2.6 THEOREM

If $M_{n+1} = \mathcal{A}M_n$ then M_{n+1} is a matroid.

Proof. We show that the set $M_{n+1} = (S \cup e_{n+1}, B_{n+1})$ satisfies **BM-1** and **BM-2**.

BM-1 $B_{n+1} \neq \emptyset$

From **AO-1** we know $|B_{n+1}| = |B_n|$. However, because M_n is a matroid, we know $|B_n| \neq 0$ from **BM-1**. Thus, $|B_{n+1}| \neq 0$ which implies $B_{n+1} \neq \emptyset$.

BM-2 For every distinct $B_1, B_2 \in B_{n+1}$, and every element $b \in B_1 - B_2$, there exists an element $\bar{b} \in B_2 - B_1$ such that $(B_1 - b) \cup \bar{b} \in B_{n+1}$

From **AO-2**, every bases in B_{n+1} can be written as the union of a basis in B_n with the element e_{n+1}. Thus, $B_1 = \bar{B}_1 \cup e_{n+1}$ and $B_2 = \bar{B}_2 \cup e_{n+1}$, where $\bar{B}_1, \bar{B}_2 \in B_n$.

Since $b \in B_1 - B_2$, and since $e_{n+1} \in B_1$ and $e_{n+1} \in B_2$, $e_{n+1} \notin B_1 - B_2$. In this case, $B_1 - B_2 = \bar{B}_1 - \bar{B}_2$. Similarly, $B_2 - B_1 = \bar{B}_2 - \bar{B}_1$.

However, since M_n is a matroid, B_n satisfies **BM-2**. Thus, there exists some element $b \in \bar{B}_1 - \bar{B}_2$, there exists an element $\bar{b} \in \bar{B}_2 - \bar{B}_1$ such that $(\bar{B}_1 - b) \cup \bar{b} \in B_n$. But b and $\bar{b}$ also satisfy **BM-2** for B_{n+1}.

$M_{n+1} = (S \cup e_{n+1}, B_{n+1})$ satisfies both **BM-1** and **BM-2**, so M_{n+1} is a matroid. ■

7.2.7 DEFINITION: EXTENSION OPERATOR

Let $\mathfrak{M}_n$ be the set of matroids on n elements with ground set $S = \{e_1, e_2, \ldots, e_n\}$. Let $\mathcal{E}: \mathfrak{M}_n \to \mathfrak{M}_{n+1}$ be an operator acting on an element $M_n \in \mathfrak{M}_n$ with bases B_n such that $\mathcal{E}M_n = M_{n+1}$ where $M_{n+1} \in \mathfrak{M}_{n+1}$ and M_{n+1} is the matroid where the bases B_{n+1} satisfy

AO-1 $|B_{n+1}| = |B_n|$

AO-2 For every $b_{n+1} \in B_{n+1}$ there exists some $b_n \in B_n$ such that $b_{n+1} = b_n$

The extension operator creates a new matroid by adding an additional element e_{n+1} to the ground set but leaving the basis unchanged.

If $M_{n+1} = \mathcal{E}M_n$ then M_{n+1} is a matroid.

Proof. The set $M_{n+1} = (S \cup e_{n+1}, B_{n+1}) = (S \cup e_{n+1}, B_n)$ satisfies **BM-1** and **BM-2** because B_n satisfies both **BM-1** and **BM-2** since M_n is a matroid. ■

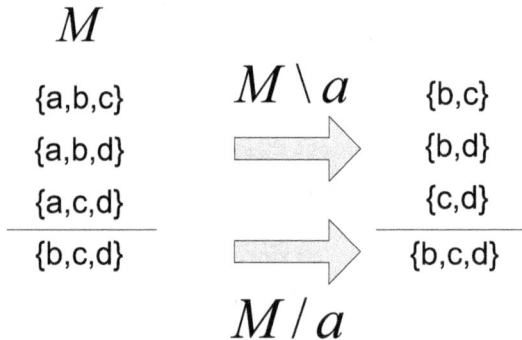

$$M$$

$\{a,b,c\}$	$M \setminus a$	$\{b,c\}$
$\{a,b,d\}$		$\{b,d\}$
$\{a,c,d\}$		$\{c,d\}$
$\{b,c,d\}$		$\{b,c,d\}$

$$M / a$$

Figure 33: Deletion and contraction of an element from a matroid.

7.2.8 EXAMPLE

Apply the augmentation operator to the matroid $M = (S, B)$ with ground set $S = \{a, b, c\}$ and set of bases $B = \{\{a, b\}\}$.

$\mathcal{A}M = \bar{M}$ where $\bar{M}$ is the matroid $\bar{M} = (\bar{S}, \bar{B})$ with ground set $S = \{a, b, c, d\}$ and set of bases $B = \{\{a, b, c\}\}$.

7.2.9 EXAMPLE

Apply the extension operator to the matroid $M = (S, B)$ with ground set $S = \{a, b, c\}$ and set of bases $B = \{\{a, b\}\}$.

$\mathcal{E}M = \bar{M}$ where $\bar{M}$ is the matroid $\bar{M} = (\bar{S}, \bar{B})$ with ground set $S = \{a, b, c, d\}$ and set of bases $B = \{\{a, b\}\}$.

7.2.10 THEOREM

Let M be a matroid with ground set S, and let M be neither extended nor augmented. Then $M = (\mathcal{E}M_1) \otimes (\mathcal{A}M_2)$ for some matroids M_1 and M_2.

Proof. Choose some element e of the ground set. If M is neither extended nor augmented, then every element of the ground set must be present in some basis, but no element is present in every basis.

Examine the deletion $M \backslash e$ as shown in Figure 33. Deletion divides the bases of M into two parts: those bases that contain e and those that do not. The bases that contain e correspond to the deletion $M \backslash e$. Moreover, the bases that do not contain e correspond to the contraction M/e.

The augmentation of the deletion $M \backslash e$ produces a matroid whose bases are the bases of M that contain e. Furthermore, the extension of the contraction M/e produces a matroid whose bases are the bases of M that do not contain e.

Consequently, $M = (\mathcal{E}M/e) \otimes (\mathcal{A}M\backslash e)$. ∎

7.2.11 THEOREM

The augmentation operator and extension operator commute so that

$$\mathcal{A}\mathcal{E}M = \mathcal{E}\mathcal{A}M$$

Proof. Examine effect of each operator on the bases of M. Let $M = (S, B)$ be a matroid on a ground set S with set of bases B. Let $|S| = n$.

The extension operator acting on M adds a new element e_{n+1} to S and leaves B unchanged. Thus,

$$\mathcal{E}M = \mathcal{E}(S, B) = (S \cup e_{n+1}, B).$$

The augmentation operator acting on M adds a new element e_{n+1} to S and adds the new element to every basis in B. Designate the new set of bases as B'.

$$\mathcal{A}M = \mathcal{A}(S, B) = (S \cup e_{n+1}, B').$$

Applying the augmentation operator to the extension of M,

$$\mathcal{A}\mathcal{E}M = \mathcal{A}\mathcal{E}(S, B) = \mathcal{A}(S \cup e_{n+1}, B) = (S \cup e_{n+1} \cup e_{n+2}, B').$$

Alternatively,

$$\mathcal{E}\mathcal{A}M = \mathcal{E}\mathcal{A}(S, B) = \mathcal{E}(S \cup e_{n+1}, B') = (S \cup e_{n+1} \cup e_{n+2}, B').$$

In both cases we arrive at the same matroid $(S \cup e_{n+1} \cup e_{n+2}, B')$. ∎

7.3 Enumeration

Because matroids have several axiomatic representations, there are many different methods that may be employed to enumerate matroids. In this section we will explore one such method based on the basis representation of matroids.

7.3.1   DEFINITION: MATROID SET

> The matroid set of order n is the set of all non-isomorphic matroids on a ground set of n elements. The matroid set of order n is designated as $\mathcal{M}_n$.

7.3.2   DEFINITION: AUGMENTED SET

> The augmented set of order n is the set of all non-isomorphic augmented matroids on a ground set of n elements. The augmented set of order n is designated as $\mathcal{A}_n$.

7.3.3   THEOREM

The number of non-isomorphic augmented matroids of order n is $|\mathcal{A}_n| = |\mathcal{M}_{n-1}|$ where $n \geq 1$.

Proof. We prove this in two parts. First, we will show there is an injective map $f: \mathcal{M}_{n-1} \to \mathcal{A}_n$. This means that $|\mathcal{A}_n| \geq |\mathcal{M}_{n-1}|$. Then we show that there is a injective map $g: \mathcal{A}_n \to \mathcal{M}_{n-1}$, which means that $|\mathcal{A}_n| \leq |\mathcal{M}_{n-1}|$. These two inequalities imply that $|\mathcal{A}_n| = |\mathcal{M}_{n-1}|$.

Let $f: \mathcal{M}_{n-1} \to \mathcal{A}_n$ be a map from the augmentation operator such that $f \circ \mathcal{M} = \mathcal{A}\mathcal{M}$ for $\mathcal{M} \in \mathcal{M}_{n-1}$. Examine the operation of f on two distinct elements $\mathcal{M}_1, \mathcal{M}_2 \in \mathcal{M}_{n-1}$. Because $\mathcal{M}_{n-1}$ is the set of non-isomorphic matroids on $n-1$ elements, $\mathcal{M}_1$ and $\mathcal{M}_2$ must be non-isomorphic since they are distinct in $\mathcal{M}_{n-1}$.

If $\mathcal{M}_1$ and $\mathcal{M}_2$ are non-isomorphic, then there is no permutation on $n-1$ elements such that $\mathcal{M}_1 = P^{(n-1)}\mathcal{M}_2$ where $P^{(n-1)}$ is a permutation on $n-1$ elements.

Let $\mathcal{A}_1 = \mathcal{A}\mathcal{M}_1$ and $\mathcal{A}_2 = \mathcal{A}\mathcal{M}_2$. Choose any presentation $\tilde{\mathcal{A}}_2$ of $\mathcal{A}_2$ such that $\tilde{\mathcal{A}}_2$ is augmented in e_n. From this initial presentation of $\mathcal{A}_2$, we may obtain all distinct presentations of $\mathcal{A}_2$ augmented in e_n by applying the permutation $P^{(n-1)}$ (we use $P^{(n-1)}$ in order to leave the presentations augmented in e_n).

Next, find a presentation of $\tilde{\mathcal{A}}_1$ of $\mathcal{A}_1$ such that $\tilde{\mathcal{A}}_1$ is augmented in e_n. If $\mathcal{A}_1$ and $\mathcal{A}_2$ are isomorphic, then there must be some permutation $P_k^{(n-1)}$ such that $\tilde{\mathcal{A}}_1 = P_k^{(n-1)} \tilde{\mathcal{A}}_2$.

But since $\mathcal{A}_2 = \mathcal{A}\mathcal{M}_2$, there must be some $\tilde{\mathcal{M}}_2$ such that $\tilde{\mathcal{A}}_2 = \mathcal{A}\tilde{\mathcal{M}}_2$. Moreover, since the augmentation operator adds the element e_n to every basis in $\tilde{\mathcal{M}}_2$, the permutation $P_k^{(n-1)}$ does not affect the augmentation, so $P_k^{(n-1)}\big(\mathcal{A}\tilde{\mathcal{M}}_2\big) = \mathcal{A}\big(P_k^{(n-1)}\tilde{\mathcal{M}}_2\big)$.

Therefore, if $\mathcal{A}_1$ and $\mathcal{A}_2$ are isomorphic then

$$\tilde{\mathcal{A}}_1 = P_k^{(n-1)} \tilde{\mathcal{A}}_2$$

$$\mathcal{A}\tilde{\mathcal{M}}_1 = P_k^{(n-1)} \mathcal{A}\tilde{\mathcal{M}}_2$$

$$\mathcal{A}\tilde{\mathcal{M}}_1 = \mathcal{A}\big(P_k^{(n-1)}\tilde{\mathcal{M}}_2\big)$$

If $\mathcal{A}_1$ and $\mathcal{A}_2$ are isomorphic, then there must be some permutation $P_k^{(n-1)}$ such that $\tilde{\mathcal{M}}_1 = P_k^{(n-1)}\tilde{\mathcal{M}}_2$. But in this case, $\mathcal{M}_1$ and $\mathcal{M}_2$ are isomorphic, in contradiction to the initial assumption that $\mathcal{M}_1$ and $\mathcal{M}_2$ are distinct members of $\mathcal{M}_{n-1}$.

From this we see that if $\mathcal{M}_1$ and $\mathcal{M}_2$ are non-isomorphic, then there augmentations $\mathcal{A}_1$ and $\mathcal{A}_2$ are also non-isomorphic. Consequently, the operator $\mathcal{A}$ is an injection from the set $\mathcal{M}_{n-1}$ to the set $\mathcal{A}_n$.

Next, start from an augmented matroid $\tilde{\mathcal{A}} \in \mathcal{A}_n$. Choose a presentation $\tilde{\mathcal{A}}$ is augmented in e_n. Let $g\colon \mathcal{A}_n \to \mathcal{M}_{n-1}$ be a map based on the deletion operator such that $g \circ \tilde{\mathcal{A}} = \tilde{\mathcal{A}}\backslash e_n$.

Examine the deletion on two distinct members $\tilde{\mathcal{A}}_1, \tilde{\mathcal{A}}_2 \in \mathcal{A}_n$ where both $\tilde{\mathcal{A}}_1$ and $\tilde{\mathcal{A}}_2$ are augmented in e_n. The result of the deletion operator is a matroid on $n-1$ elements. Let $\tilde{\mathcal{M}}_1 = \tilde{\mathcal{A}}_1\backslash e_n$ and $\tilde{\mathcal{M}}_2 = \tilde{\mathcal{A}}_2\backslash e_n$ where $\tilde{\mathcal{M}}_1, \tilde{\mathcal{M}}_2 \in \mathcal{M}_{n-1}$.

Since $\tilde{\mathcal{A}}_1$ and $\tilde{\mathcal{A}}_2$ are non-isomorphic, there is no permutation $P_k^{(n)}$ such that $\tilde{\mathcal{A}}_1 = P_k^{(n)}\tilde{\mathcal{A}}_2$. Moreover, if there is no permutation of n elements with this property, then there is no permutation on $n-1$ elements either.

Furthermore, because the permutation on $n-1$ elements does not affect e_n, it does not affect the action deletion. Thus, $P_k^{(n-1)}(\tilde{\mathcal{A}}_2 \setminus e_n) = (P_k^{(n-1)}\tilde{\mathcal{A}}_2)\setminus e_n$.

If $\tilde{\mathcal{M}}_1$ and $\tilde{\mathcal{M}}_2$ are isomorphic, then there must be a permutation such that

$$\tilde{\mathcal{M}}_1 = P_k^{(n-1)}\tilde{\mathcal{M}}_2$$

$$\tilde{\mathcal{A}}_1 \setminus e_n = P_k^{(n-1)}(\tilde{\mathcal{A}}_2 \setminus e_n)$$

$$\tilde{\mathcal{A}}_1 \setminus e_n = \left(P_k^{(n-1)}\tilde{\mathcal{A}}_2\right)\setminus e_n$$

However, since no permutation exists where $\tilde{\mathcal{A}}_1 = P_k^{(n-1)}\tilde{\mathcal{A}}_2$, then $\tilde{\mathcal{M}}_1$ and $\tilde{\mathcal{M}}_2$ must be non-isomorphic. Consequently, the map $g: \mathcal{A}_n \to \mathcal{M}_{n-1}$ where g is the deletion operator such that $g \circ \tilde{\mathcal{A}} = \tilde{\mathcal{A}}\setminus e_n$ is injective.

If there is an injective map $f: \mathcal{M}_{n-1} \to \mathcal{A}_n$ that maps each element of $\mathcal{M}_{n-1}$ to a unique element of $\mathcal{A}_n$, it must be true that $|\mathcal{A}_n| \geq |\mathcal{M}_{n-1}|$. Furthermore, since $g: \mathcal{A}_n \to \mathcal{M}_{n-1}$ is also injective, it must be the case that $|\mathcal{M}_{n-1}| \geq |\mathcal{A}_n|$. Therefore, $|\mathcal{A}_n| = |\mathcal{M}_{n-1}|$. ∎

7.3.4 DEFINITION: EXTENDED SET

> The extended set of order n is the set of all non-isomorphic extended matroids on a ground set of n elements. The extended set of order n is designated as $\mathcal{E}_n$.

7.3.5 THEOREM

The number of non-isomorphic extended matroids of order n is $|\mathcal{E}_n| = |\mathcal{M}_{n-1}|$ where $n \geq 1$.

Proof. The proof follows the proof of the previous theorem, replacing the augmentation operator $\mathcal{A}$ with the extension operator $\mathcal{E}$, and substituting contraction for deletion. ∎

7.3.6 THEOREM

The sets $\mathcal{A}_n$ and $\mathcal{E}_n$ are dual to each other.

Proof. $\mathcal{E}_n$ is the set of all extended matroids on the ground set S with n elements, while $\mathcal{A}_n$ is the set of all augmented matroids on S. Let $\tilde{A} \in \mathcal{A}_n$ be an augmented matroid on S. There must be some element e that is present in every basis of $\tilde{A}$.

The dual to $\tilde{A}$ is the matroid $\tilde{A}^*$ where every basis b^* in $\tilde{A}^*$ is constructed from some basis b in $\tilde{A}$ as $b^* = S - b$. However, since e is in every b, e will not be in any b^*. This means that the matroid $\tilde{A}^*$ is extended in e, so $\tilde{A}^* \in \mathcal{E}_n$. From this, the dual forms an injective map $*: \mathcal{A}_n \to \mathcal{E}_n$.

Applying the dual to an element $\tilde{E} \in \mathcal{E}_n$, since there is some e that is not present in every basis b of $\tilde{E}$, e must be present in each basis b^* of $\tilde{E}^*$. In this case the dual forms an injective map $*: \mathcal{E}_n \to \mathcal{A}_n$.

Finally, since the dual is invertible (the dual of the dual is the original matroid), these maps are bijective. The dual of element $\tilde{A} \in \mathcal{A}_n$ maps uniquely to an element $\tilde{E} \in \mathcal{E}_n$ and vice versa. ∎

7.3.7 THEOREM

The number of non-isomorphic matroids $\mathcal{I}_n$ of order n that are both augmented and extended is $|\mathcal{I}_n| = |\mathcal{M}_{n-2}|$ where $n \geq 2$.

Proof. From the previous proofs we see that the set of augmented and extended matroids may be produced through the action of $\mathcal{A}$ or $\mathcal{E}$ on the set of non-isomorphic matroids $\mathcal{M}_{n-1}$. If a matroid is both augmented and extended, then is must result from the action of both $\mathcal{A}$ and $\mathcal{E}$.

Let $\mathcal{I}_n$ be the set of matroids on n elements that are both augmented and extended. If $\tilde{J} \in \mathcal{I}_n$, then $\tilde{J} = \mathcal{A}\mathcal{E}\tilde{M} = \mathcal{E}\mathcal{A}\tilde{M}$, where $\tilde{M} \in \mathcal{M}_{n-2}$.

Since the map $\mathcal{A}: \mathcal{M}_{n-1} \rightarrow \mathcal{A}_n$ is injective and $\mathcal{A}_n \subset \mathcal{M}_n$, $\mathcal{A}$ is also injective as a map $\mathcal{A}: \mathcal{M}_{n-1} \rightarrow \mathcal{M}_n$. Similarly, the map $\mathcal{E}: \mathcal{M}_{n-1} \rightarrow \mathcal{E}_n$ is injective, and so is $\mathcal{E}: \mathcal{M}_{n-1} \rightarrow \mathcal{M}_n$.

Consequently, the compound map $\mathcal{E}\mathcal{A}: \mathcal{M}_{n-2} \rightarrow \mathcal{M}_n$ is also injective. Furthermore, the matroids obtained from the action of both augmentation and extension are the members of the set $\mathcal{I}_n$. Thus, the compound map is effectively $\mathcal{E}\mathcal{A}: \mathcal{M}_{n-2} \rightarrow \mathcal{I}_n$. But since this map is injective then $|\mathcal{I}_n| \geq |\mathcal{M}_{n-2}|$.

From the previous theorems we know that both the map $\backslash e_n: \mathcal{A}_n \rightarrow \mathcal{M}_{n-1}$ and $/e_n: \mathcal{E}_n \rightarrow \mathcal{M}_{n-1}$ are injective. Since $\mathcal{I}_n \subset \mathcal{A}_n$ the map $\backslash e_n: \mathcal{I}_n \rightarrow \mathcal{M}_{n-1}$ is also injective. Also, if we delete e_n from an extended matroid, the resulting matroid is still extended.

Combining these results we have $\backslash e_n: \mathcal{I}_n \rightarrow \mathcal{E}_{n-1}$. If we apply contraction to the an element of $\mathcal{E}_{n-1}$ that is extended in e_{n-1} we have the map $/e_{n-1}: \mathcal{E}_{n-1} \rightarrow \mathcal{M}_{n-2}$. Both of these maps are injective, so the combined map $/e_{n-1}\backslash e_n: \mathcal{I}_n \rightarrow \mathcal{M}_{n-2}$ is injective.

But if this map is injective, then it must be true that $|\mathcal{M}_{n-2}| \geq |\mathcal{I}_n|$. Combining this with the earlier inequality where $|\mathcal{I}_n| \geq |\mathcal{M}_{n-2}|$ we have $|\mathcal{I}_n| = |\mathcal{M}_{n-2}|$. ∎

7.3.8 THEOREM

Let M be a matroid with ground set S on n elements. If $M = (\mathcal{E}M_1) \otimes (\mathcal{A}\mathcal{A}M_2)$ for some matroids M_1 and M_2, then M is augmented.

Proof. The set of bases B of M are created through the union of the bases of $\mathcal{E}M_1$ and $\mathcal{A}\mathcal{A}M_2$. Let $B_\mathcal{E}$ be the set of bases of $\mathcal{E}M_1$ and $B_\mathcal{A}$ be the set of bases of $\mathcal{A}\mathcal{A}M_2$.

Each $b_\mathcal{A} \in B_\mathcal{A}$ contains the elements e_n and e_{n-1} because $\mathcal{A}\mathcal{A}M_2$ is double augmented. Since M is a matroid, we know from Matroid Product theorem 1.1.2 for every $b_\mathcal{A} \in B_\mathcal{A}$ and $b_\mathcal{E} \in B_\mathcal{E}$, every element

$b \in b_{\mathcal{A}} - b_{\mathcal{E}}$, there exists an element $\bar{b} \in b_{\mathcal{E}} - b_{\mathcal{A}}$ such that $(b_{\mathcal{A}} - b) \cup \bar{b} \in B$.

Since $\mathcal{E}M_1$, there is no $b_{\mathcal{E}} \in B_{\mathcal{E}}$ such that $e_n \in b_{\mathcal{E}}$. Suppose there is some $b_{\mathcal{E}} \in B_{\mathcal{E}}$ such that $e_{n-1} \notin b_{\mathcal{E}}$. Then $e_n, e_{n-1} \in b_{\mathcal{A}} - b_{\mathcal{E}}$. If we choose $b = e_{n-1}$ we find that $(b_{\mathcal{A}} - e_{n-1}) \cup \bar{b} \in B$. However, since $e_n \in b_{\mathcal{A}}$, the basis $(b_{\mathcal{A}} - e_{n-1}) \cup \bar{b}$ contains e_n but does not contain e_{n-1}. This is impossible because no basis of $\mathcal{E}M_1$ contains e_n, and every basis of $\mathcal{A}\mathcal{A}M_2$ has both e_n and e_{n-1}.

From this, it must be the case that there is no $b_{\mathcal{E}} \in B_{\mathcal{E}}$ such that $e_{n-1} \notin b_{\mathcal{E}}$. But then $\mathcal{E}M_1$ is augmented in e_{n-1}. Moreover, since $\mathcal{A}\mathcal{A}M_2$ is also augmented in e_{n-1}, so the product $M = (\mathcal{E}M_1) \otimes (\mathcal{A}\mathcal{A}M_2)$ is augmented in e_{n-1}.∎

7.3.9 THEOREM

Let M be a matroid with ground set S on n elements. If $M = (\mathcal{E}\mathcal{E}M_1) \otimes (\mathcal{A}M_2)$ for some matroids M_1 and M_2, then M is extended.

Proof. This proof follows the previous proof. Let $B_{\mathcal{E}}$ be the set of bases of $\mathcal{E}\mathcal{E}M_1$ and $B_{\mathcal{A}}$ be the set of bases of $\mathcal{A}M_2$.

Each $b_{\mathcal{E}} \in B_{\mathcal{E}}$ does not have either e_n or e_{n-1} because $\mathcal{E}\mathcal{E}M_1$ is double extended. But since M is a matroid, the Matroid Product theorem 1.1.2 states for every $b_{\mathcal{E}} \in B_{\mathcal{E}}$ and $b_{\mathcal{A}} \in B_{\mathcal{A}}$, every element $b \in b_{\mathcal{A}} - b_{\mathcal{E}}$, there exists an element $\bar{b} \in b_{\mathcal{E}} - b_{\mathcal{A}}$ such that $(b_{\mathcal{E}} - b) \cup \bar{b} \in B$.

Since $\mathcal{A}M_2$, every $b_{\mathcal{A}} \in B_{\mathcal{A}}$ contains e_n. Suppose there is some $b_{\mathcal{A}} \in B_{\mathcal{A}}$ such that $e_{n-1} \in b_{\mathcal{A}}$. Then $e_n, e_{n-1} \in b_{\mathcal{A}} - b_{\mathcal{E}}$. If we choose $b = e_{n-1}$ we find that $(b_{\mathcal{A}} - e_{n-1}) \cup \bar{b} \in B$. However, since $e_n \in b_{\mathcal{A}}$, the basis $(b_{\mathcal{A}} - e_{n-1}) \cup \bar{b}$ contains e_n but does not contain e_{n-1}. This is impossible because no basis of $\mathcal{E}\mathcal{E}M_1$ contains e_n, and every basis of $\mathcal{A}M_2$ has both e_n and e_{n-1}.

From this, it must be the case that there is no $b_{\mathcal{A}} \in B_{\mathcal{A}}$ such that $e_{n-1} \in b_{\mathcal{A}}$. But then $\mathcal{A}M_2$ is extended in e_{n-1}. Again, since $\mathcal{E}\mathcal{E}M_1$ is also extended in e_{n-1}, so the product $M = (\mathcal{E}\mathcal{E}M_1) \otimes (\mathcal{A}M_2)$ is extended in e_{n-1}.∎

7.3.10   THEOREM

Every non-isomorphic matroid on n elements may be constructed from one of the following processes:

$$\mathcal{E}\mathcal{M}_{n-1}$$
$$\mathcal{A}\mathcal{M}_{n-1}$$
$$\mathcal{E}_n^1 \otimes \mathcal{A}_n^1$$

where $\mathcal{M}_{n-1}$ is the set of non-isomorphic matroids on $n-1$ elements, $\mathcal{E}_n^1$ is the set of singly extended matroids on n elements, and $\mathcal{A}_n^1$ is the set of singly augmented matroids on n elements.

Proof. From theorem 7.2.10, every matroid is either augmented, extended, or from a product $(\mathcal{E}M_1) \otimes (\mathcal{A}M_2)$.

Examining the product, theorem 7.3.9 states if M_1 is extended, then the product $(\mathcal{E}M_1) \otimes (\mathcal{A}M_2)$ will produce an extended matroid. But all extended matroids can be constructed from $\mathcal{E}\mathcal{M}_{n-1}$. Because of this, we only need to be concerned with M_1's that are not extended. In this case, $\mathcal{E}M_1$ is just the set of singly extended matroids on n elements $\mathcal{E}_n^1 \subseteq \mathcal{E}_n$.

Similarly, if M_2 is augmented, theorem 7.3.8 insists that the product will produce an augmented matroid, which is already produced in the set of augmented matroids $\mathcal{A}_n = \mathcal{A}\mathcal{M}_{n-1}$. Consequently, we only need to examine the set of singly augmented matroids $\mathcal{A}_n^1 \subseteq \mathcal{A}_n$.

Based on the above arguments, all matroids on n elements may be constructed from $\mathcal{E}\mathcal{M}_{n-1}$, $\mathcal{A}\mathcal{M}_{n-1}$, and products of the form $\mathcal{E}_n^1 \otimes \mathcal{A}_n^1$. ∎

7.3.11   THEOREM

The number of non-isomorphic singly extended matroids of order n is $|\mathcal{E}_n^1| = |\mathcal{M}_{n-1}| - |\mathcal{M}_{n-2}|$ where $n \geq 2$. The number of non-isomorphic singly augmented matroids of order n is $|\mathcal{A}_n^1| = |\mathcal{M}_{n-1}| - |\mathcal{M}_{n-2}|$ where $n \geq 2$.

Proof. The set of all extended matroids of order n is constructed from $\mathcal{E}\mathcal{M}_{n-1}$. Every matroid that is more than singly extended is constructed by applying the extension operator at least twice. Specifically, the set $\mathcal{E}\mathcal{E}\mathcal{M}_{n-2}$ contains all matroids on n elements that are at least twice extended. Since application of the extension operator produces a unique matroid, the total number of singly extended matroids is $|\mathcal{E}_n^1| = |\mathcal{M}_{n-1}| - |\mathcal{M}_{n-2}|$.

A similar argument can be made for singly augmented matroids. This leads to the similar result $|\mathcal{A}_n^1| = |\mathcal{M}_{n-1}| - |\mathcal{M}_{n-2}|$. ∎

Chapter 8: Generalizations of Matroids

There are many ways to generalize the basic matroid concept. In this chapter, we explore several matroid generalizations and discuss some of their properties.

8.1 Signed Sets

Matroids are mathematical structures over a finite set. We can exend this to similar structures defined over a finite signed set. This section introduces the concepts of signed sets and some of their properties.

8.1.1 DEFINITION: SIGNED SET

> Let S be a finite set. A signed set $\bar{S}$ is a partition of S into two disjoint sets $\hat{S} = (S^+, S^-)$, where $S^+ \cup S^- = S$, and $S^+ \cap S^- = \emptyset$.

There are three main methods used to denote a signed set. Each method proves useful in certain situations. The sections below explain the notation and provide examples.

Signed Set Notation

The signed set notation explicitly lists the elements of S^+ and S^-. As an example let $S = \{a, b, c, d, e, f, g\}$ and let the positive partition of S be $S^+ = \{a, c, d, g\}$ and the negative partition $S^- = \{b, e, f\}$.

First, note that the partition satisfies the required properties:

$$S^+ \cup S^- = \{a, b, c, d, e, f, g\} = S$$
$$S^+ \cap S^- = \emptyset$$

The signed set notation for this partition is

$$\hat{S} = (S^+, S^-) = (\{a, c, d, g\}, \{b, e, f\}).$$

Bar Notation

In bar notation, the set elements are listed as an ordered string where a bar is placed over the negative elements. Using the example from above,

$$\hat{S} = a\bar{b}cd\bar{e}\bar{f}g.$$

Signed Incidence Notation

The signed incidence notation lists a signed value for every element of S in order. The signed value is:

$$v_i = \begin{cases} + & \textit{Element is positive} \\ - & \textit{Element is negative}. \\ 0 & \textit{Element not present} \end{cases}$$

The previous example is written in signed incidence notation as

$$\hat{S} = + - + + - -.$$

8.1.2 DEFINITION: SIGNED SET COMPLEMENT

Let $\hat{S} = (S^+, S^-)$ be a signed set. The complement of $\hat{S}$, denoted $-\hat{S}$, is the signed set $-\hat{S} = (S^-, S^+)$.

The complement of a signed set is also called a reorientation.

8.1.3 EXAMPLE

Let $S = \{a, b, c, d\}$ be a set partitioned into the sets $S^- = \{a, b\}$ and $S^+ = \{c, d\}$. Find the signed set, bar, and signed incidence notations for the signed set and its complement.

The table below provides the notations for the signed set and its complement.

	Signed Set	Bar	Signed Incidence
$\hat{S}$	$(\{a, b\}, \{c, d\})$	$a b \bar{c} \bar{d}$	$+ + - -$
$-\hat{S}$	$(\{c, d\}, \{a, b\})$	$\bar{a} \bar{b} c d$	$- - + +$

8.1.4 EXAMPLE

Let $S = \{a, b, c, d, e, f\}$ be a set and let $S' = \{a, b, d, f\}$ be a subset of S. Suppose S' is partitioned into the sets $S^+ = \{a, f\}$ and $S^+ = \{b, d\}$. Find the signed set, bar, and signed incidence notations for the signed set and its complement.

The following table provides the notation for the signed set and its reorientation:

	Signed Set	Bar	Signed Incidence
$\hat{S}$	$(\{a, f\}, \{b, d\})$	$a\bar{b}\bar{d}f$	$+ - 0 - 0 +$
$-\hat{S}$	$(\{b, d\}, \{a, f\})$	$\bar{a}bd\bar{f}$	$- + 0 + 0 -$

In this example, it is important to note that we are examining the signed set on S', not on S. Because of this we have $S^+ \cup S^- = S'$, instead of $S^+ \cup S^- = S$.

This example also demonstrates how the signed incidence notation treats missing elements. The signed incidence notation provides an entry for every element in the superset S. If the element is not present in the signed subset under consideration, the signed incidence notation enters a 0 at the location for the missing element.

8.1.5 DEFINITION: UNDERLYING SET

Let $\hat{S} = (S^+, S^-)$ be a signed set. The underlying set $\underline{S}$ is the union of the positive and negative sets of $\hat{S}$.

$$\underline{S} = S^+ \cup S^-.$$

The underlying set connected a signed set back to an ordinary unsigned set.

8.1.6   DEFINITION: SIGNED SET COMPOSITION

Let $\hat{X} = (X^+, X^-)$ and $\hat{Y} = (Y^+, Y^-)$ be signed sets. The composition operator $\circ$ produces a new signed set $\hat{Z} = \hat{X} \circ \hat{Y} = (Z^+, Z^-)$ where

$$Z^+ = X^+ \cup (Y^+ - X^-)$$
$$Z^- = X^- \cup (Y^- - X^+).$$

The signed set composition operator provides a means to create a new signed set from two other signed sets.

8.1.7   EXAMPLE

Let $\hat{X} = (\{a, b\}, \{c, d\})$ and let $\hat{Y} = (\{a\}, \{b\})$ be signed sets. Find the composition $\hat{Z}_1 = \hat{X} \circ \hat{Y}$ and $\hat{Z}_2 = \hat{Y} \circ \hat{X}$.

For $\hat{Z}_1 = \hat{X} \circ \hat{Y}$ we have

$$Z_1^+ = X^+ \cup (Y^+ - X^-) = \{a, b\} \cup (\{a\} - \{c, d\}) = \{a, b\} \cup \{a\} = \{a, b\}$$
$$Z_1^- = X^- \cup (Y^- - X^+) = \{c, d\} \cup (\{b\} - \{a, b\}) = \{c, d\} \cup \emptyset = \{c, d\}$$

For $\hat{Z}_2 = \hat{Y} \circ \hat{X}$ we have

$$Z_2^+ = Y^+ \cup (X^+ - Y^-) = \{a\} \cup (\{a, b\} - \{b\}) = \{a\} \cup \{a\} = \{a\}$$
$$Z_2^- = Y^- \cup (X^- - Y^+) = \{b\} \cup (\{c, d\} - \{a\}) = \{b\} \cup \{c, d\} = \{b, c, d\}$$

8.1.8   EXAMPLE

Let $\hat{X} = (X^+, X^-)$ be signed set. Find the composition $\hat{Z}_1 = \hat{X} \circ \hat{X}$ and $\hat{Z}_2 = \hat{X} \circ (-\hat{X})$.

Because $X^+ \cap X^- = \emptyset$, $X^+ - X^- = X^+$ and $X^- - X^+ = X^-$. Then for $\hat{Z}_1 = \hat{X} \circ \hat{X}$ we have

$$Z_1^+ = X^+ \cup (X^+ - X^-) = X^+ \cup X^+ = X^+$$
$$Z_1^- = X^+ \cup (X^- - X^+) = X^- \cup X^- = X^-$$

so we find $\hat{Z}_1 = \hat{X} \circ \hat{X} = \hat{X}$.

Similarly for $\hat{Z}_2 = \hat{X} \circ (-\hat{X})$,

$$Z_2^+ = X^+ \cup (X^- - X^-) = X^+ \cup \emptyset = X^+$$
$$Z_2^- = X^- \cup (X^+ - X^+) = X^- \cup \emptyset = X^-$$

Here we find $\hat{Z}_2 = \hat{X} \circ (-\hat{X}) = \hat{X}$.

8.1.9 THEOREM

The signed set composition operation is not necessarily commutative.

Proof. We have already provided an explicit example demonstrating that the signed set composition operation is not commutative. We will show that this is the case on more general grounds by sketching the general result using Venn diagrams.

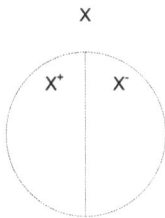

Figure 34: Signed sets X and Y in isolation.

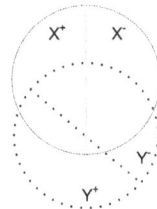

Figure 35: Sets X and Y with overlapping elements.

Figure 36: The positive set of $\hat{X} \circ \hat{Y}$.

Figure 37: The negative set of $\hat{X} \circ \hat{Y}$.

Figure 38: The positive set of $\hat{Y} \circ \hat{X}$.

Figure 39: The negative set of $\hat{Y} \circ \hat{X}$.

Figure 34-Figure 39 show Venn diagrams for the positive and negative sets of $\hat{Z}_1 = \hat{X} \circ \hat{Y}$ and $\hat{Z}_2 = \hat{Y} \circ \hat{X}$. The initial sets $\hat{X}$ and $\hat{Y}$ are first shown in isolation in Figure 34. Figure 35 displays the signed sets $\hat{X}$ and $\hat{Y}$ with their overlapping elements. In the general case, the composition is taken between sets where some elements of $\hat{X}$ are not present in $\hat{Y}$ and vice-versa.

Figure 36 represents the portion of the overlap that makes up the positive portion of the signed set $\hat{Z}_1 = \hat{X} \circ \hat{Y}$, while Figure 37 shows the negative portion. It is clear from this set that if the positive portion is united with the negative portion we have the entire original set.

Figure 38 displays the positive portion of $\hat{Z}_2 = \hat{Y} \circ \hat{X}$, while Figure 39 shows the negative set. It is clear that these sets are distinct from the positive and negative portions of $\hat{Z}_1 = \hat{X} \circ \hat{Y}$. ∎

Figure 41: Sets X and Y with overlapping elements.

Figure 40: Signed sets X and Y in isolation.

Figure 42: The positive set of $\hat{X} \circ \hat{Y}$.

Figure 43: The negative set of $\hat{X} \circ \hat{Y}$.

Figure 44: The positive set of $\hat{Y} \circ \hat{X}$.

Figure 45: The negative set of $\hat{Y} \circ \hat{X}$.

8.1.10 THEOREM

Let $\hat{X} = (X^+, X^-)$ and $\hat{Y} = (Y^+, Y^-)$ be signed sets where $\underline{\hat{X}} \subseteq \underline{\hat{Y}}$. Then $\hat{Z}_1 = \hat{X} \circ \hat{Y} \supseteq \hat{X}$ but $\hat{Z}_2 = \hat{Y} \circ \hat{X} = \hat{Y}$.

Proof. Again, we sketch a proof by examining Venn diagrams for the sets. Figure 40 shows the sets $\hat{X}$ and $\hat{Y}$ in isolation, while Figure 41 presents the sets with overlapping elements. We can see in Figure 41 that $\hat{X} \subseteq \hat{Y}$.

Figure 42 displays the positive set of $\hat{X} \circ \hat{Y}$. This set contains the entire portion of X^+. Moreover, Figure 43 shows the negative portion of $\hat{X} \circ \hat{Y}$. Again, we see that $\hat{Z}_1^- \supseteq X^-$. From these diagrams we see that the composition $\hat{Z}_1 = \hat{X} \circ \hat{Y} \supseteq \hat{X}$ is a superset of $\hat{X}$ in the sense that

$$\hat{Z}_1^+ \supseteq X^+ \quad \hat{Z}_1^- \supseteq X^-.$$

Figure 44 shows the positive portion of $\hat{Z}_2 = \hat{Y} \circ \hat{X}$. In this case, the positive portion of $\hat{Z}_2$ is simply the positive portion of $\hat{Y}$. Similarly, the negative portion of $\hat{Z}_2$ is the negative portion of $\hat{Y}$. ∎

8.2   Oriented Matroids

Oriented matroids extend the matroid concepts already discussed to signed sets. Oriented matroids can be used to examine digraphs and ordered vector spaces. Many of the concepts discussed earlier can be extended to oriented matroids. However, a full examination of these concepts is outside the scope of this book.

8.2.1   DEFINITION: ORIENTED CIRCUIT MATROID

Let S be a finite set. Let $\hat{C}$ be a collection of signed subsets of S with the following properties:

COM-1 $\hat{\emptyset} \notin \hat{C}$

COM-2 If $\hat{c}_1, \hat{c}_2 \in \hat{C}$ and $\underline{c}_1 \subseteq \underline{c}_2$, then $\hat{c}_1 = \hat{c}_2$ or $\hat{c}_1 = -\hat{c}_2$

COM-3 If $\hat{c}_1, \hat{c}_2$ are distinct members of $\hat{C}$ where $\hat{c}_1 \neq -\hat{c}_2$, and $e \in \hat{c}_1^+ \cap \hat{c}_2^-$, then there exists a $\hat{c}_3 \in \hat{C}$ such that $\hat{c}_3^+ \subseteq (\hat{c}_1^+ \cup \hat{c}_2^+) - e$ and $\hat{c}_3^- \subseteq (\hat{c}_1^- \cup \hat{c}_2^-) - e$

COM-4 $\hat{C} = -\hat{C}$

Then the system $(S, \hat{C})$ is an oriented matroid.

The similarity between the definition of a circuit matroid and an oriented circuit matroid is readily apparent. We will explore these similarities later in this section.

Furthermore, as circuit matroids often arise from the consideration of graphs, oriented circuit matroids arise from the consideration of digraphs. We will explore this through examples of oriented matroids from digraphs.

In this section we will use the hat notation $\hat{}$ to differentiate signed sets versus ordinary sets. In the above definition, $\hat{C}, \hat{c}_1, \hat{c}_2, \hat{c}_3$ are all

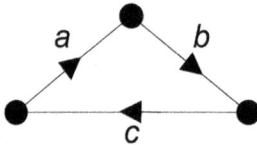

Figure 46: Simple 3-vertex diagraph with an oriented cycle.

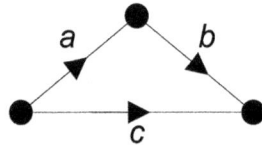

Figure 47: Simple 3-vertex diagraph without an oriented cycle.

signed sets, while S, c_1 and c_2 are ordinary sets. Furthermore e is an element of a set and does not have an associated sign.

8.2.2 EXAMPLE

Find the oriented matroid from the cycles of the digraph in Figure 46.

The ground set S is the set of edges $S = \{a, b, c\}$. The only cycle is the set $\{a, b, c\}$. To determine an orientation for the cycle edges, we must fix a reference frame to the graph. When examining a cycle, we choose to call edges pointing in the clockwise direction as positive, and edges pointing in the counterclockwise direction negative.

Using this choice of reference, the signed cycle set is

$$\hat{C} = \{(\{a, b, c\}, \emptyset)\}.$$

However, **COM-4** insists that $\hat{C} = -\hat{C}$. In order for this to hold, we need to add the reoriented cycles into the cycle set. With this, the complete signed cycle set is

$$\hat{C} = \{(\{a, b, c\}, \emptyset), (\emptyset, \{a, b, c\})\}.$$

The oriented matroid is the system $\hat{M} = (S, \hat{C})$. This example has an oriented cycle. An oriented cycle is a cycle where all of the edges point in the same direction. We can recognize oriented cycles in an oriented matroid because these cycles have either $c^+ = \emptyset$ or $c^- = \emptyset$.

8.2.3 EXAMPLE

Find the oriented matroid from the cycles of the digraph in Figure 47.

The ground set S is again the set of edges $S = \{a, b, c\}$. Similarly, the only cycle is the set $\{a, b, c\}$. We choose the same orientation as in the previous example, so

$$\hat{C} = \{(\{a, b\}, \{b\})\}.$$

In this case there is no oriented cycle so neither $c^+ = \emptyset$ nor $c^- = \emptyset$. The oriented matroid is the system $\hat{M} = (S, \hat{C})$.

8.2.4 DEFINITION: ORIENTED VECTOR MATROID

Let S be a signed set. Let $\hat{V}$ be a collection of signed subsets of S with the following properties:

VOM-1 $\hat{\emptyset} \notin \hat{V}$

VOM-2 If $\hat{v}_1, \hat{v}_2 \in \hat{V}$ then $\hat{v}_1 \circ \hat{v}_2 \in \hat{V}$

VOM-3 If $\hat{v}_1, \hat{v}_2 \in \hat{V}$ are distinct members of $\hat{C}$ where $\hat{v}_1 \neq -\hat{v}_2$,
$$e \in \hat{v}_1{}^+ \cap \hat{v}_2{}^-, \quad \text{and} \quad f \in \left(\hat{v}_1 - \hat{v}_2\right) \cup \left(\hat{v}_2 - \hat{v}_1\right) \cup$$
$(\hat{v}_1{}^+ \cap \hat{v}_2{}^+) \cup (\hat{v}_1{}^- \cap \hat{v}_2{}^-)$, then there exists a $v_3 \in \hat{V}$ such that $\hat{v}_3{}^+ \subseteq (\hat{v}_1{}^+ \cup \hat{v}_2{}^+) - e$, $\hat{v}_3{}^- \subseteq (\hat{v}_1{}^- \cup \hat{v}_2{}^-) - e$, and $f \in \underline{\hat{V}}$.

VOM-4 $\hat{V} = -\hat{V}$

Then the system $(S, \hat{C})$ is an oriented matroid.

This is the oriented matroid equivalent of a vector matroid. The axioms for an oriented vector matroid are similar to the axioms of an ordinary vector matroid.

In the case of vector matroids, the independent sets are sets of vectors that are linearly independent,

$$\sum_k a_k v_k = 0.$$

One perspective to obtain these independent sets is to first find all of the minimal dependent sets, then the independent sets are the proper subsets of the dependent sets (and we need to add $\emptyset$ to the list of independent sets as well).

This approach is better suited to understanding the signed aspects of the sets in a vector matroid. First, identify the minimal dependent sets satisfying

$$\sum_k a_k v_k = 0.$$

We can create a signed set from this by assigning the v_k's into either a positive or negative set depending on the sign of a_k. In this manner we can obtain a collection of signed sets. The independent sets are these sets and all their subsets.

8.2.5 EXAMPLE

Find the oriented vector matroid from the 2×3 matrix $\begin{bmatrix} 1 & 0 & 1 \\ 0 & 1 & 1 \end{bmatrix}$.

Label the columns a, b and c respectively; $S = \{a, b, c\}$ with corresponding vectors $v_a = \begin{pmatrix} 1 \\ 0 \end{pmatrix}$, $v_b = \begin{pmatrix} 0 \\ 1 \end{pmatrix}$ and $v_c = \begin{pmatrix} 1 \\ 1 \end{pmatrix}$. The minimal dependent set is

$$v_a + v_b - v_c = 0.$$

We create the signed set $\hat{C} = (\{v_a, v_b\}, \{v_c\})$. The independent signed subsets are:

ab	$a\bar{c}$	$b\bar{c}$
a	b	$\bar{c}$
	$\emptyset$	

From **VOM-4**, $\hat{V} = -\hat{V}$ so we need to extend this set by adding its reorientation. The reorientations are

$\bar{a}\bar{b}$	$\bar{a}c$	$\bar{b}c$
$\bar{a}$	$\bar{b}$	c

8.3 Greedoids

Greedoids are a generalization of matroids designed to capture some aspects of the optimality of the greedy algorithm. Greedoids are also used to model formal languages.

8.3.1 DEFINITION: GREEDOID

Let S be a finite set. Let $\mathfrak{G}$ be a collection of subsets of S with the following properties:

GD-1 For every $G \in \mathfrak{G}$, $G \neq \emptyset$, there exists a $g \in G$ such that $G - g \in \mathfrak{G}$

GD-2 If $G_1, G_2 \in \mathfrak{G}$ with $|G_1| > |G_2|$, there exists a $g \in G_1 - G_2$ such that $G_2 \cup g \in \mathfrak{G}$

Then the ordered pair $(S, \mathfrak{G})$ is called a greedoid.

A greedoid is a generalization of a matroid. There is no requirement that $\emptyset$ be a member. In fact, there is no requirement that every subset of an independent set (feasible set) in a greedoid be part of the greedoid. Instead, if G is an independent set of the greedoid, we must also have some subset of G be in the greedoid where the subset has cardinality one less than G.

8.3.2 EXAMPLE

Let S and $\mathfrak{G}$ be

$$S = \{a, b\}$$
$$\mathfrak{G} = \{\emptyset, \{a\}, \{a, b\}\}$$

Show that $(S, \mathfrak{G})$ is a greedoid.

We need to show that $(S, \mathfrak{G})$ satisfies **GD-1** and **GD-2**.

GD-1 For every $G \in \mathfrak{G}$, $G \neq \emptyset$, there exists a $g \in G$ such that
$$G - g \in \mathfrak{G}$$

There are three sets in $\mathfrak{G}$, let $G_1 = \{a, b\}$, $G_2 = \{a\}$, and $G_3 = \emptyset$. Examine each separately.

$G_1 = \{a, b\}$

We need to identify an element $g \in G_1$ such that $G_1 - g \in \mathfrak{G}$. If we set $g = b$, we have $G_1 - g = \{a\} = G_2 \in \mathfrak{G}$.

$G_2 = \{a\}$

Again, we need to identify an element $g \in G_2$ such that $G_1 - g \in \mathfrak{G}$. If we set $g = a$, we have $G_2 - g = \emptyset = G_3 \in \mathfrak{G}$.

$G_3 = \emptyset$

In this case, $G_3 = \emptyset$, so **GD-1** does not apply.

We see that in each case, **GD-1** is satisfied.

GD-2 If $G_1, G_2 \in \mathfrak{G}$ with $|G_1| > |G_2|$, there exists a $g \in G_1 - G_2$ such that $G_2 \cup g \in \mathfrak{G}$

There are three pairs to examine. Let $G_1 = \{a, b\}$, $G_2 = \{a\}$, and $G_3 = \emptyset$. We examine each pair of sets and show that **GD-2** is satisfied.

G_1, G_2

Let $D = G_1 - G_2 = \{b\}$. The union $G_2 \cup \{b\} = \{a, b\} = G_1 \in \mathfrak{G}$.

G_1, G_3

Let $D = G_1 - G_3 = \{a, b\}$. Choose $g = \{a\}$. The union $G_3 \cup \{a\} = \{a\} = G_2 \in \mathfrak{G}$.

G_2, G_3

Let $D = G_2 - G_3 = \{a\}$. The union $G_3 \cup \{a\} = \{a\} = G_2 \in \mathfrak{G}$.

Thus, **GD-1** and **GD-2** are satisfied. Then the system $(S, \mathfrak{G})$ is a greedoid.

8.3.3   1.1.3   THEOREM

Let M be a matroid. Then M is also a greedoid.

Proof. Because M is a matroid, **SD-1**, **SD-2**, and **SD-3** are satisfied. **SD-3** is equivalent to **GD-2**. Furthermore, **GD-1** follows immediately from **SD-2**. Thus, M satisfies **GD-1** and **GD-2** so M is a greedoid. ∎

8.3.4   DEFINITION: GREEDOID BASIS

> Let G be a greedoid with ground set S and feasible set $\mathfrak{G}$. A basis of G is a maximal set in $\mathfrak{G}$.

A basis of a greedoid is defined in the same manner as for a matroid.

8.3.5   DEFINITION: GREEDOID RANK

> Let G be a greedoid with ground set S and feasible set $\mathfrak{G}$. The rank of G is the cardinality of any basis of G.

The rank of a greedoid is defined similarly to a matroid. The exchange property **GD-2** for a greedoid is equivalent to **SD-3** which leads to the conclusion that all bases have the same cardinality.

8.4   Antimatroids

Antimatroids are a specialization of greedoids in the same sense that matroids are a specialization of greedoids. Antimatroids are a particular type of interval greedoid.

8.4.1   DEFINITION: INTERVAL GREEDOID

> Let G be a greedoid with ground set S and feasible set $\mathfrak{G}$.
>
> **IG-1**   Let $G_1, G_2, G_3 \in \mathfrak{G}$ such that $G_1 \subseteq G_2 \subseteq G_3$. Let $g \in S - G_3$. If $G_3 \cup g \in \mathfrak{G}$ and $G_1 \cup g \in \mathfrak{G}$ then $G_2 \cup g \in \mathfrak{G}$
>
> Then G is an interval greedoid.

The property **IG-1** is called the interval property.

8.4.2 EXAMPLE

Let S and $\mathfrak{G}$ be

$$S = \{a, b, c\}$$
$$\mathfrak{G} = \{\emptyset, \{a\}, \{a, b\}\}$$

Show that $(S, \mathfrak{G})$ is an interval greedoid.

We need to show that $(S, \mathfrak{G})$ satisfies **GD-1**, **GD-2**, and **IG-1**.

GD-1 For every $G \in \mathfrak{G}$, $G \neq \emptyset$, there exists a $g \in G$ such that $G - g \in \mathfrak{G}$

There are three sets in $\mathfrak{G}$, let $G_1 = \{a, b\}$, $G_2 = \{a\}$, and $G_3 = \emptyset$. Examine each separately.

$G_1 = \{a, b\}$

We need to identify an element $g \in G_1$ such that $G_1 - g \in \mathfrak{G}$. If we set $g = b$, we have $G_1 - g = \{a\} = G_2 \in \mathfrak{G}$.

$G_2 = \{a\}$

Again, we need to identify an element $g \in G_2$ such that $G_1 - g \in \mathfrak{G}$. If we set $g = a$, we have $G_2 - g = \emptyset = G_3 \in \mathfrak{G}$.

$G_3 = \emptyset$

In this case, $G_3 = \emptyset$, so **GD-1** does not apply.

We see that in each case, **GD-1** is satisfied.

GD-2 If $G_1, G_2 \in \mathfrak{G}$ with $|G_1| > |G_2|$, there exists a $g \in G_1 - G_2$ such that $G_2 \cup g \in \mathfrak{G}$

There are three pairs to examine. Let $G_1 = \{a, b\}$, $G_2 = \{a\}$, and $G_3 = \emptyset$. We examine each pair of sets and show that **GD-2** is satisfied.

G_1, G_2

Let $D = G_1 - G_2 = \{b\}$. The union $G_2 \cup \{b\} = \{a, b\} = G_1 \in \mathfrak{G}$.

G_1, G_3

Let $D = G_1 - G_3 = \{a, b\}$. Choose $g = a$. The union $G_3 \cup \{a\} = \{a\} = G_2 \in \mathfrak{G}$.

G_2, G_3

Let $D = G_2 - G_3 = \{a\}$. The union $G_3 \cup \{a\} = \{a\} = G_2 \in \mathfrak{G}$.

Thus, **GD-1** and **GD-2** are satisfied. Then the system $(S, \mathfrak{G})$ is a greedoid.

IG-1 Let $G_1, G_2, G_3 \in \mathfrak{G}$ such that $G_1 \subseteq G_2 \subseteq G_3$. Let $g \in S - G_3$. If $G_3 \cup g \in \mathfrak{G}$ and $G_1 \cup g \in \mathfrak{G}$ then $G_2 \cup g \in \mathfrak{G}$.

There are only three sets in $\mathfrak{G}$. Again, set $G_1 = \{a, b\}$, $G_2 = \{a\}$, and $G_3 = \emptyset$. Let $D = S - G_1 = \{c\}$. Since D has only one element, we set $g = c$.

Now, $G_3 \cup g = \{a, b, c\} \notin \mathfrak{G}$. There is no set $G_3 \cup g \in \mathfrak{G}$, so **IG-1** is trivially satisfied.

8.4.3 DEFINITION: ANTIMATROID

> Let G be an interval greedoid with ground set S and feasible set $\mathfrak{G}$.
>
> **AM-1** Let $G_1, G_2 \in \mathfrak{G}$ such that $G_1 \subseteq G_2$. Let $g \in S - G_2$. If $G_1 \cup g \in \mathfrak{G}$ then $G_2 \cup g \in \mathfrak{G}$

An antimatroid a specialization of an interval greedoid.

8.5 Lagrangian Matroids

Lagrangian matroids are another matroid generalization. We will define Lagrangian matroids by generalizing the matroid basis definition. Recall that a basis matroid is defined in terms of a ground set and a collection of subsets. Specifically, let S be a set and let $\mathfrak{B}$ be a collection of subsets of S where $\mathfrak{B}$ satisfies

BM-1 $\mathfrak{B} \neq \emptyset$

BM-2 For every distinct $A, B \in \mathfrak{B}$, and every element $a \in A - B$, there exists an element $b \in B - A$ such that $(A - a) \cup b \in \mathfrak{B}$

Let $\mathfrak{I}$ be the collection of all subsets of the elements of $\mathfrak{B}$. Then the ordered pair $(S, \mathfrak{I})$ is a matroid.

8.5.1 DEFINITION: BASIS MATROID

Let S be a set and let $\mathfrak{B}$ be a collection of subsets of S where $\mathfrak{B}$ satisfies

LBM-1 $\mathfrak{B} \neq \emptyset$

LBM-2 For every distinct $A, B \in \mathfrak{B}$, and every element $a \in A - B$, there exists an element $b \in B - A$ such that $A \triangle \{a, b\} \in \mathfrak{B}$

Let $\mathfrak{I}$ be the collection of all subsets of the elements of $\mathfrak{B}$. Then the ordered pair $(S, \mathfrak{I})$ is a matroid.

We see this is identical to the original definition of a basis matroid, with the exception that the statement $(A - a) \cup b \in \mathfrak{B}$ is replaces with the symmetric difference $A \triangle \{a, b\} \in \mathfrak{B}$. We begin by showing that these statements are equivalent.

From the definition of symmetric difference,

$$A \triangle \{a, b\} = (A \cup \{a, b\}) - (A \cap \{a, b\})$$

Now, let $c \in A$ such that $c \neq a$. We know that $c \neq b$ because $b \in B - A$ which means that $b \notin A$. Thus, $c \in A \cup \{a, b\}$ but $c \notin A \cap \{a, b\}$. Thus, $c \in A \triangle \{a, b\}$. Moreover, $b \in A \cup \{a, b\}$ but $b \notin A \cap \{a, b\}$. This means that $b \in A \triangle \{a, b\}$. Finally, $a \in A \cup \{a, b\}$ and $a \in A \cap \{a, b\}$, so $a \notin A \triangle \{a, b\}$.

From this, we see that the set $A \triangle \{a, b\}$ contains all members of A except a, and the additional element b. Thus,

$$A \triangle \{a, b\} = (A - a) \cup b.$$

8.5.2   DEFINITION: Δ-MATROID

Let S be a set and let $\mathfrak{B}$ be a collection of subsets of S where $\mathfrak{B}$ satisfies

LSM-1 $\mathfrak{B} \neq \emptyset$

LSM-2 For every distinct $A, B \in \mathfrak{B}$, and every element $a \in A \triangle B$, there exists an element $b \in B \triangle A$ such that $A \triangle \{a, b\} \in \mathfrak{B}$

The ordered pair $(S, \mathfrak{B})$ is a Δ-matroid.

The Δ-matroid extends the definition of a matroid by allowing $a \in A \triangle B$ rather than just $a \in A - B$. Similarly, $b \in B \triangle A$ instead of $b \in B - A$. Because $A - B \in A \triangle B$ and $B - A \in B \triangle A$, we see that this generalizes the definition of a matroid.

In addition, **LSM-2** is not as restrictive as **LBM-2**. While **LBM-2** forces all elements of $\mathfrak{B}$ to be equicardinal, **LSM-2** does not. Furthermore, the set of feasible (independent) sets for the Δ-matroid do not need to contain every subset of an element $b \in \mathfrak{B}$.

8.5.3   EXAMPLE

Let $S = \{a, b, c\}$ and let $\mathfrak{B} = \{\{a, b, c\}, \{c\}\}$. Show that the ordered pair $(S, \mathfrak{B})$ is a Δ-matroid.

We only need to show that $\mathfrak{B}$ satisfies **LSM-1** and **LSM-2**. $\mathfrak{B}$ clearly satisfies **LSM-1** since $\mathfrak{B}$ is not empty. As for **LSM-2**, we need to examine each element of the symmetric difference $a \in A \triangle B$ and identify an element $b \in B \triangle A$ such that $A \triangle \{a, b\} \in \mathfrak{B}$.

We examine the symmetric difference of every distinct pair of elements in $\mathfrak{B}$.

$A = \{a, b\}$ $B = \{a\}$

First, find the symmetric difference:

$$A \triangle B = B \triangle A = \{a, b\} \triangle \{a\} = \{a, b\} \cup \{a\} - \{a, b\} \cap \{a\} = \{b\}$$

There is only one element to choose. We test the symmetric difference:

$$A \Delta \{b\} = \{a\} \quad B \Delta \{b\} = \{a, b\}$$

The reader can verify that the remaining pairs satisfy **LSM-2**.

8.5.4 DEFINITION: TWISTED MATROID

Let $M = (S, \mathfrak{I})$ be a matroid (or Δ-matroid) with independent sets $\mathfrak{I}$. Let T be a set, not necessarily a subset of S. The twisting of M with respect to T is the ordered pair $(S \cup T, \mathfrak{I}_T)$ where

$$\mathfrak{I}_T = \{I \Delta T \mid I \in \mathfrak{I}\}$$

is a twisted matroid.

A twisted matroid is a Δ-matroid. We sketch a proof using the Venn diagrams in Figure 48-Figure 53.

Figure 48 shows two representative sets, I_1, I_2, along with the twist T. The symmetric differences $I_1 \Delta T$ and $I_2 \Delta T$ are showin in Figure 49 and Figure 50 respectively.

We need to find the symmetric difference between the sets of Figure 49 and Figure 50. The symmetric difference is written in terms of a union and an intersection. Figure 51 shows the union while Figure 52 provides the intersection.

From this, the symmetric difference is the difference of the set of Figure 51 and Figure 52. The set representing the symmetric difference is depicted in Figure 53. From this we see that the symmetric difference between two elements in $\mathfrak{I}_I$ of a twist matroid is identical to the symmetric difference between the sets of underlying matroid.

This result, coupled with definition **LBM-2**, shows that **LSM-2** is satisfied.

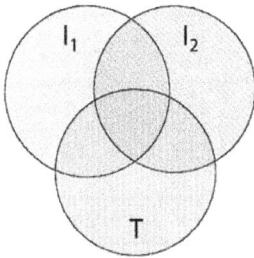

Figure 48: Venn diagram of the sets I_1, I_2, and T.

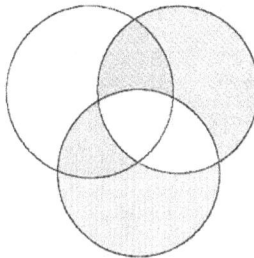

Figure 49: $I_1 \Delta T$.

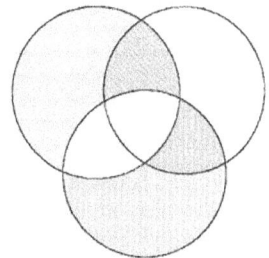

Figure 50: $I_2 \Delta T$.

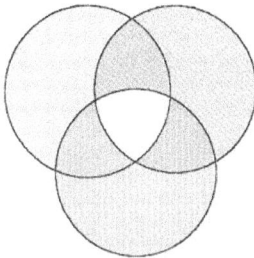

Figure 51: $(I_1 \Delta T) \cup (I_2 \Delta T)$.

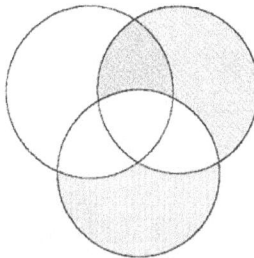

Figure 52: $(I_1 \Delta T) \cap (I_2 \Delta T)$.

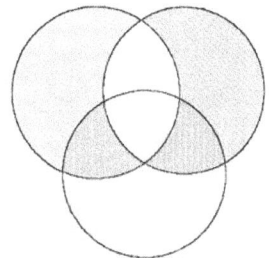

Figure 53: $(I_1 \Delta T) \Delta (I_2 \Delta T) = I_1 \Delta I_2$.

8.5.5   DEFINITION: SYMMETRIC SET

Let $S = \{a, b, c, \ldots n\}$ be a set. Let $S^* = \{a^*, b^*, c^*, \ldots n^*\}$ be a different set such that $S \cap S^* = \emptyset$ and $|S| = |S^*|$, where the elements of S and S^* are related by an operator with the properties

$$\bar{a}_i = a_i^* \quad \bar{a}_i^* = a_i$$

The union $\Sigma = S \cup S^*$ is called a symmetric set.

Symmetric sets are useful in defining symmetric and symplectic matroids. We also require a definition for admissible sets provided next.

8.5.6   DEFINITION: ADMISSIBLE SET

Let $\Sigma = S \cup S^*$ be a symmetric set. Let $A \subseteq \Sigma$ be a subset. Define the set $\bar{A} = \{\bar{a}_i \mid a_i \in A\}$. A is an admissible set if $A \cap \bar{A} = \emptyset$.

8.5.7 DEFINITION: SYMMETRIC MATROID

Let Σ be a symmetric set and let $\mathcal{B}$ be a collection of subsets of Σ where $\mathcal{B}$ satisfies

LYM-1 $\mathcal{B} \neq \emptyset$

LYM-2 For every distinct $A, B \in \mathcal{B}$, and every element $a \in A \Delta B$, there exists an element $b \in B \Delta A$ such that $A \Delta \{a, b, a^*, b^*\} \in \mathcal{B}$

The ordered pair $(\Sigma, \mathcal{B})$ is a symmetric matroid.

8.5.8 DEFINITION: GALE ORDERING

Let $\Sigma = \{\sigma_1, \sigma_2, ..., \sigma_{2n}\}$ be a symmetric set and $A, B \subseteq \Sigma$ be sequences of elements of Σ. Define a partial ordering $\prec$ on Σ. Let $A = \{a_1, a_2, ..., a_k\}$ be a sequence where $a_i \prec a_{i+1}$ for every element in A. Similarly, let $B = \{b_1, b_2, ..., b_k\}$ be a sequence where $b_i \prec b_{i+1}$ for every element in B. The sequences A and B are Gale ordered if $a_i \prec b_i$ for $1 \leq i \leq k$.

8.5.9 EXAMPLE

Let $\Sigma = \{a, b, c, d\}$ be a symmetric ordered set obeying the order relation

$$a \prec b \quad b \prec c \quad c \prec d.$$

Show that the sequences $A = \{a, c\}$ and $B = \{b, d\}$ are Gale ordered.

We examine each element of the sequences. We see that $a \prec b$ and $c \prec d$, so A and B are Gale ordered.

8.5.10 DEFINITION: B_n-ADMISSIBLE ORDERING

Let $\Sigma = \{\sigma_1, \sigma_2, ..., \sigma_n, \sigma_1^*, \sigma_2^*, ..., \sigma_n^*\}$ be a symmetric set with a total ordering $\prec$. An admissible sequence on Σ is B_n-admissible if and only if

$$\sigma_1 \prec \sigma_2 \prec \cdots \prec \sigma_n \prec \sigma_n^* \prec \sigma_{n-1}^* \prec \cdots \prec \sigma_1^*.$$

We require a total ordering on Σ in order to produce a B_n-admissible sequence.

8.5.11 EXAMPLE

Let $\Sigma = \{a, b, c, d\}$ be a symmetric totally ordered set obeying the order relation

$$a \prec b \quad b \prec c \quad c \prec d,$$

where

$$a^* = d \quad b^* = c.$$

Show that the sequence $A = \{a, c\}$ is B_n-admissible.

First, we show that A is admissible. Construct the set $\bar{A} = \{\bar{a}_i \mid a_i \in A\}$. We have $\bar{A} = \{\bar{a}, \bar{c}\} = \{d, b\}$. The intersection $A \cap \bar{A} = \emptyset$, so A is an admissible sequence.

Next, we show that the sequence A satisfies the total order relation. In fact, since $a \prec b$ and $b \prec c$, from the transitive property of the ordering relation we know that $a \prec c$, so A is B_n-admissible.

8.5.12 DEFINITION: D_n-ADMISSIBLE ORDERING

Let $\Sigma = \{\sigma_1, \sigma_2, ..., \sigma_n, \sigma_1^*, \sigma_2^*, ..., \sigma_n^*\}$ be a symmetric set with a partial ordering $\prec$. An admissible sequence on Σ is D_n-admissible if and only if

$$\sigma_1 \prec \sigma_2 \prec \cdots \prec \frac{\sigma_n}{\sigma_n^*} \prec \sigma_{n-1}^* \prec \cdots \prec \sigma_1^*.$$

We require a partial ordering on Σ in order to produce a D_n-admissible sequence. The partial ordering must have two chains and only a single pair of incomparable elements.

8.5.13 EXAMPLE

Let $\Sigma = \{a, b, c, d\}$ be a symmetric totally ordered set obeying the order relation

$$a \prec b \quad c \prec d,$$

where

$$a^* = d \quad b^* = c.$$

Show that the sequence $A = \{a, c\}$ is D_n-admissible.

Similar to the previous example, we know that A is admissible because $\bar{A} = \{\bar{a}, \bar{c}\} = \{d, b\}$, and the intersection $A \cap \bar{A} = \emptyset$.

Furthermore, the sequence A satisfies the partial order relation. Because $a \prec b$ and $c \prec d$ we see that A is D_n-admissible.

8.5.14 DEFINITION: CLASSICAL MATROID

Let S be a set, $\mathfrak{I}_k$ the collection all k-element subsets of S, and $\mathfrak{B} \subseteq \mathfrak{I}_k$. The set $(S, \mathfrak{I}_k)$ is a classical matroid if and only if for every linearly ordering $\prec$ on S there exists some $B \in \mathfrak{B}$ such that $A \prec B$ for every $A \in \mathfrak{B}, A \neq B$.

A classical matroid has a linear ordering with a maximal element.

8.5.15 DEFINITION: SYMPLECTIC MATROID

Let S be a symmetric set, $\mathfrak{I}_k$ the collection all k-element B_n-admissible subsets of S, and $\mathfrak{B} \subseteq \mathfrak{I}_k$. The set $(S, \mathfrak{I}_k)$ is a symplectic matroid if and only if for every linearly ordering $\prec$ on S there exists some $B \in \mathfrak{B}$ such that $A \prec B$ for every $A \in \mathfrak{B}, A \neq B$.

A symplectic matroid has a total ordering with a maximal element.

8.5.16 DEFINITION: ORTHOGONAL MATROID

Let S be a symmetric set, $\mathfrak{I}_k$ the collection all k-element D_n-admissible subsets of S, and $\mathfrak{B} \subseteq \mathfrak{I}_k$. The set $(S, \mathfrak{I}_k)$ is an orthogonal matroid if and only if for every linearly ordering $\prec$ on S there exists some $B \in \mathfrak{B}$ such that $A \prec B$ for every $A \in \mathfrak{B}, A \neq B$.

An orthogonal matroid has a partial ordering with a maximal element.

8.5.17 DEFINITION: LAGRANGIAN MATROID

> Let $\mathfrak{M}$ be a symplectic or orthogonal matroid with ground set S. $\mathfrak{M}$ is Lagrangian if $\mathfrak{M}$ has maximal rank $k = n = |S|/2$.

8.5.18 EXAMPLE

Let $\Sigma = \{a, b, c, d\}$ be a symmetric ordered set obeying the order relation

$$a \prec b \quad b \prec c \quad c \prec d,$$

where

$$a^* = d \quad b^* = c.$$

Let $\mathfrak{B} = \{\{a, b\}, \{c, d\}\}$. Show that the ordered pair $(\Sigma, \mathfrak{B})$ is a Lagrangian symplectic matroid.

The ordering $\prec$ on Σ is a total ordering because each pair of element is comparable, either directly or through the transitive ordering property.

To find the admissible collection $\mathfrak{I}_2$, we need to check that each element $J \in \mathfrak{I}_2$ satisfies $J \cap \bar{J} = \emptyset$.

$$\{a, b\} \cap \overline{\{a, b\}} = \{a, b\} \cap \{d, c\} = \emptyset \quad \{a, c\} \cap \overline{\{a, c\}} = \{a, c\} \cap \{d, b\} = \emptyset$$
$$\{b, d\} \cap \overline{\{b, d\}} = \{b, d\} \cap \{c, a\} = \emptyset \quad \{c, d\} \cap \overline{\{c, d\}} = \{c, d\} \cap \{b, a\} = \emptyset$$

These sets form the collection $\mathfrak{I}_2$ and obey a total order relation. Thus, $\mathfrak{I}_2 = \{\{a, b\}, \{a, c\}, \{b, d\}, \{c, d\}\}$ is B_n-admissible. Moreover, we see that $\mathfrak{B} \subseteq \mathfrak{I}_2$.

Next, we note that every element of $\mathfrak{B}$ is $\prec$ than $\{c, d\}$. Since there are only two elements, we only need to observe that $\{a, b\} \prec \{c, d\}$. Thus, the ordered pair $(\Sigma, \mathfrak{B})$ is a symplectic matroid.

Finally, we see that the cardinality of the elements of $\mathfrak{B}$ is exactly $n = 2 = |\Sigma|/2$. Thus, the ordered pair $(\Sigma, \mathfrak{B})$ is a Lagrangian symplectic matroid.

Answers to Problems

1.1 SETS

1. a. 3
   b. 6
   c. 2
2. $S = \{a, e, i, o, u\}$ cardinality = 5

1.2 SUBSETS

1. $\emptyset$ - Improper
   $\{a\}$ - Proper
   $\{b\}$ - Proper
   $\{a, b\}$ – Improper

2.
   a. $S \cup T = \{-1,0,1,2\}$
   b. $S \cap T = \{0,1\}$
   c. $S - T = \{-1\}$
   d. $T - S = \{2\}$
   e. $S \Delta T = \{-1,2\}$

3.
   a. $S \cup T = \{a, b, c, d\}$
   b. $S \cap T = \{b, c\}$
   c. $S - T = \{a, d\}$
   d. $T - S = \emptyset$
   e. $S \Delta T = \{a, d\}$

1.3 MAXIMAL AND MINIMAL SETS

1. $\lceil S \rceil = \{\{a, b\}, \{b, c\}\}, \lfloor S \rfloor = \{\{a\}, \{b\}, \{c\}\}$

1.5 FUNCTIONS

1. Surjection
2. Injection
3. Bijection

1.8 Transversals

1. There are three transversals of this system:
   a. $T = \{a, d\}$
   b. $T = \{b, d\}$
   c. $T = \{c, d\}$

2. There are no transversals of this system: $T = \emptyset$

3. There are five partial transversals:
   a. $PT = \{a\}$
   b. $PT = \{b\}$
   c. $PT = \{c\}$
   d. $PT = \{d\}$
   e. $PT = \emptyset$

4. There are six partial transversals:
   a. $PT = \{a\}$
   b. $PT = \{b\}$
   c. $PT = \{c\}$
   d. $PT = \{a, c\}$
   e. $PT = \{b, c\}$
   f. $PT = \emptyset$

2.1 Vector Matroids

7. $v_a = \begin{pmatrix} 1 \\ 0 \end{pmatrix}, v_b = \begin{pmatrix} 0 \\ 1 \end{pmatrix}, v_c = \begin{pmatrix} 1 \\ 1 \end{pmatrix}, v_d = \begin{pmatrix} 0 \\ 1 \end{pmatrix}$

$I = \{\emptyset, \{a\}, \{b\}, \{c\}, \{d\}, \{a, b\}, \{a, c\}, \{a, d\}, \{b, c\}, \{c, d\}\}$

8. $v_a = \begin{pmatrix} 1 \\ 0 \\ 0 \end{pmatrix}, v_b = \begin{pmatrix} 0 \\ 1 \\ 0 \end{pmatrix}, v_c = \begin{pmatrix} 0 \\ 0 \\ 1 \end{pmatrix}, v_d = \begin{pmatrix} 0 \\ 1 \\ 1 \end{pmatrix}, v_e = \begin{pmatrix} 1 \\ 1 \\ 0 \end{pmatrix}$

$I = \left\{ \begin{array}{c} \emptyset, \{a\}, \{b\}, \{c\}, \{d\}, \{e\}, \{a, b\}, \{a, c\}, \{a, d\}, \{a, e\}, \\ \{b, c\}, \{b, d\}, \{b, e\}, \{c, d\}, \{c, e\}, \{d, e\}, \\ \{a, b, c\}, \{a, b, d\}, \{a, c, d\}, \{a, c, e\}, \{a, d, e\}, \{b, c, e\}, \{b, d, e\}, \{c, d, e\} \end{array} \right\}$

9. $R = \{\emptyset, \{a\}, \{b\}, \{a, b\}\}$

10. $R = \{\emptyset, \{a\}, \{b\}, \{c\}, \{a, b\}, \{a, c\}, \{b, c\}, \{a, b, c\}\}$

3.1 RANK

1. 3
2.
    a. 2
    b. 3
    c. 2
    d. 1
3.
    a. 1
    b. 0
    c. 0
    d. 0
4. 2
5.
    a. 1
    b. 2
    c. 2
    d. 1
    e. 1
    f. 0
6.
    a. 2
    b. 1
    c. 0
    d. 1
    e. 0
    f. 1

3.2 CLOSURE

1.
    a. $\{a, b, c, d, e\}$
    b. $\{a, b, c, d, e\}$
    c. $\{a, e\}$
    d. $\{a, b, c, d, e\}$
    e. $\{b, e\}$
    f. $\{e\}$
    g. $\{a, b, c\}$

   h. $\{a, b, c, d\}$
   i. $\{a, b, c\}$
   j. $\{a, d\}$
   k. $\{b, e\}$
   l. $\{d\}$

2. $\{e\}, \{a, e\}, \{b, e\}, \{c, e\}, \{d, e\}, \{a, b, c, d, e\}$

3. $\emptyset, \{a\}, \{b\}, \{c\}, \{d\}, \{a, d\}, \{b, d\}, \{c, d\}, \{a, b, c\}, \{a, b, c, d\}$

3.3 BASIS

1. $I = \emptyset, L = \emptyset$
2. $I = \emptyset, L = \{f\}$

4.1 DELETION

1.

   a. $S = \{b, c, d\}, B = \{\{b, c\}, \{b, d\}\}$
   b. $S = \{c, d\}, B = \{\{c\}, \{d\}\}$
   c. $S = \{b, d\}, B = \{\{b, d\}\}$
   d. $S = \{a, b\}, B = \{\{a\}, \{b\}\}$

2.

   a. $S = \{a, b, c, d\}, B = \{\{a, c, d\}, \{b, c, d\}\}$
   b. $S = \{a, b, c\}, B = \{\{a, c\}, \{b, c\}\}$
   c. $S = \{a, b\}, B = \{\{a\}, \{b\}\}$
   d. $S = \{a\}, B = \{\{a\}\}$
   e. $S = \{a, b\}, B = \{\{a\}, \{b\}\}$

3.

   a. $S = \{b, c\}, B = \{\{b, c\}\}$
   b. $S = \{a, c\}, B = \{\{a, c\}\}$
   c. $S = \{a, b\}, B = \{\{a\}, \{b\}\}$
   d. $S = \{c\}, B = \{\{c\}\}$
   e. $S = \{b\}, B = \{\{b\}\}$
   f. $S = \{a\}, B = \{\{a\}\}$

4.2 DUAL

1. $S = \{a, b, c, d\}, B = \{\{a, b\}, \{a, c\}, \{b, c\}\}$
2. $S = \{a, b, c, d\}, B = \{\{a\}, \{b\}, \{c\}, \{d\}\}$

4.3 CONTRACTION

1.

    a. $S = \{b, c\}, B = \{\{c\}\}$

    b. $S = \{a, c\}, B = \{\{c\}\}$

    c. $S = \{a, b\}, B = \{\{a\}, \{b\}\}$

    d. $S = \{c\}, B = \{\{c\}\}$

    e. $S = \{b\}, B = \{\{b\}\}$

    f. $S = \{a\}, B = \{\{a\}\}$

2.

    a. $S = \{b, c\}, B = \{\emptyset\}$

    b. $S = \{a, c\}, B = \{\emptyset\}$

    c. $S = \{a, b\}, B = \{\emptyset\}$

    d. $S = \{c\}, B = \{\emptyset\}$

    e. $S = \{b\}, B = \{\emptyset\}$

    f. $S = \{a\}, B = \{\emptyset\}$

3.

    a. $S = \{a, b, d\}, B = \{\{d\}\}$

    b. $S = \{a, b, c\}, B = \{\{a\}, \{c\}\}$

4.5 SUMS

1. $S = \{a, b\}, B = \{\{a, b\}\}$

2. $S = \{a, b, c\}, B = \{\{a, b\}\}$

3. $S = \{a, b, c\}, B = \{\{a, b, c\}\}$

4. $S = \{a, b, c, d\}, B = \{\{a, b, c, d\}\}$

6.1 GREEDY ALGORITHM

1. $X_0 = \emptyset$

    $X_1 = \{c\}$

    $X_2 = \{b, c\}$

    This is the global maximum of the sets in T. The set (S, T) is a matroid.

2. $X_0 = \emptyset$

    $X_1 = \{d\}$

    This is not the global maximum of the sets in T ($T = \{b, c\}$ is the maximum). The set (S, T) is not a matroid.

Matroid Catalog

The following is a catalog of all matroids on six or fewer elements. Each matroid is listed specifying a basis, rank, number of distinct permutations, and wither it is an augmented or extended matroid. In addition, polynomials are given for the rank generating function (RG), rank polynomial (RP), characteristic polynomial (χ), and distribution polynomial (DP). Finally, the Möbius function is given, except for matroids where the Möbius function is too large to fit the page.

Matroids on One Element

Matroid: 1-1

Basis: {a}

Rank: 1   Permutations: 1   Augmented: Y   Extended: N

RG: $1 + \theta$

RP: $1 + x$

$\chi : -1 + x$

DP: $\theta\varphi$

Möbius Function:

	{}	{a}
{}	1	-1
{a}	0	1

Matroid: 1-2

Basis:

Rank: 0   Permutations: 1   Augmented: N   Extended: Y

RG: $1 + \varphi$

RP: 1

$\chi : 1$

DP: φ

Möbius Function:

	{a}
{a}	1

Matroids on Two Elements

Matroid: 2-1

Basis: {a}

Rank: 1   Permutations: 2   Augmented: Y   Extended: Y

RG: $1 + \varphi + \theta + \theta\varphi$

RP: $1 + x$

$\chi : -1 + x$

DP: $\varphi + \theta\varphi$

Möbius Function:

	{b}	{a,b}
{b}	1	-1
{a,b}	0	1

Matroid: 2-2

Basis: {a}, {b}

Rank: 1   Permutations: 1   Augmented: N   Extended: N

RG: $2 + \varphi + \theta$

RP: $1 + x$

$\chi : -1 + x$

DP: $\theta\varphi^2$

Möbius Function:

	{}	{a,b}
{}	1	-1
{a,b}	0	1

Matroid: 2-3

Basis:

Rank: 0   Permutations: 1   Augmented: N   Extended: Y

RG: $1 + 2\varphi + \varphi^2$

RP: 1

$\chi : 1$

DP: φ^2

Möbius Function:

	{a,b}
{a,b}	1

Matroid: 2-4

Basis: {a,b}

Rank: 2   Permutations: 1   Augmented: Y   Extended: N

RG: $1 + 2\theta + \theta^2$

RP: $1 + 2x + x^2$

$\chi : 1 - 2x + x^2$

DP: $\theta\varphi^2$

Möbius Function:

	{}	{a}	{b}	{a,b}
{}	1	-1	-1	1
{a}	0	1	0	-1
{b}	0	0	1	-1
{a,b}	0	0	0	1

Matroids on Three Elements

Matroid: 3-1

Basis: {a}

Rank: 1   Permutations: 3   Augmented: Y   Extended: Y

RG: $1 + 2\varphi + \varphi^2 + \theta + 2\theta\varphi + \theta\varphi^2$

RP: $1 + x$

$\chi : -1 + x$

DP: $\varphi^2 + \theta\varphi$

Möbius Function:

	{b,c}	{a,b,c}
{b,c}	1	-1
{a,b,c}	0	1

Matroid: 3-2

Basis: {a}, {b}

Rank: 1   Permutations: 3   Augmented: N   Extended: Y

RG: $2 + 3\varphi + \varphi^2 + \theta + \theta\varphi$

RP: $1 + x$

$\chi : -1 + x$

DP: $\varphi + \theta\varphi^2$

Möbius Function:

	{c}	{a,b,c}
{c}	1	-1
{a,b,c}	0	1

Matroid: 3-3

Basis: {a}, {b}, {c}

Rank: 1   Permutations: 1   Augmented: N   Extended: N

RG: $3 + 3\varphi + \varphi^2 + \theta$

RP: $1 + x$

$\chi : -1 + x$

DP: $\theta\varphi^3$

Möbius Function:

	{}	{a,b,c}
{}	1	-1
{a,b,c}	0	1

Matroid: 3-4

Basis:

Rank: 0   Permutations: 1   Augmented: N   Extended: Y

RG: $1 + 3\varphi + 3\varphi^2 + \varphi^3$

RP: 1

$\chi : 1$

DP: φ^3

Möbius Function:

	{a,b,c}
{a,b,c}	1

Matroid: 3-5

Basis: {a,b}

Rank: 2   Permutations: 3   Augmented: Y   Extended: Y

RG: $1 + \varphi + 2\theta + 2\theta\varphi + \theta^2 + \theta^{2\varphi}$

RP: $1 + 2x + x^2$

$\chi : 1 - 2x + x^2$

DP: $\varphi + \theta\varphi^2$

Möbius Function:

	{c}	{a,c}	{b,c}	{a,b,c}
{c}	1	-1	-1	1
{a,c}	0	1	0	-1
{b,c}	0	0	1	-1
{a,b,c}	0	0	0	1

Matroid: 3-6

Basis: {a,b}, {a,c}

Rank: 2   Permutations: 3   Augmented: Y   Extended: N

RG: $2 + \varphi + 3\theta + \theta\varphi + \theta^2$

RP: $1 + 2x + x^2$

$\chi: 1 - 2x + x^2$

DP: $\theta\varphi^2 + \theta^2\varphi$

Möbius Function:

	{}	{a}	{b,c}	{a,b,c}
{}	1	-1	-1	1
{a}	0	1	0	-1
{b,c}	0	0	1	-1
{a,b,c}	0	0	0	1

Matroid: 3-7

Basis: {a,b}, {a,c}, {b,c}

Rank: 2   Permutations: 1   Augmented: N   Extended: N

RG: $3 + \varphi + 3\theta + \theta^2$

RP: $1 + 3x + x^2$

$\chi: 2 - 3x + x^2$

DP: $\theta^2\varphi^3$

Möbius Function:

	{}	{a}	{b}	{c}	{a,b,c}
{}	1	-1	-1	-1	2
{a}	0	1	0	0	-1
{b}	0	0	1	0	-1
{c}	0	0	0	1	-1
{a,b,c}	0	0	0	0	1

Matroid: 3-8

Basis: {a,b,c}

Rank: 3   Permutations: 1   Augmented: Y   Extended: N

RG: $1 + 3\theta + 3\theta^2 + \theta^3$

RP: $1 + 3x + 3x^2 + x^3$

$\chi: -1 + 3x - 3x^2 + x^3$

DP: $\theta\varphi^3$

Möbius Function:

	{}	{a}	{b}	{c}	{a,b}	{a,c}	{b,c}	{a,b,c}
{}	1	-1	-1	-1	1	1	1	-1
{a}	0	1	0	0	-1	-1	0	1
{b}	0	0	1	0	-1	0	-1	1
{c}	0	0	0	1	0	-1	-1	1
{a,b}	0	0	0	0	1	0	0	-1
{a,c}	0	0	0	0	0	1	0	-1
{b,c}	0	0	0	0	0	0	1	-1
{a,b,c}	0	0	0	0	0	0	0	1

Matroids on Four Elements

Matroid: 4-1

Basis: {a}

Rank: 1 Permutations: 4 Augmented: Y Extended: Y

RG: $1 + 3\varphi + 3\varphi^2 + \varphi^3 + \theta + 3\theta\varphi + 3\theta\varphi^2 + \theta\varphi^3$

RP: $1 + x$

$\chi : -1 + x$

DP: $\varphi^3 + \theta\varphi$

Möbius Function:

	{b,c,d}	{a,b,c,d}
{b,c,d}	1	-1
{a,b,c,d}	0	1

Matroid: 4-2

Basis: {a}, {b}

Rank: 1 Permutations: 6 Augmented: N Extended: Y

RG: $2 + 5\varphi + 4\varphi^2 + \varphi^3 + \theta + 2\theta\varphi + \theta\varphi^2$

RP: $1 + x$

$\chi : -1 + x$

DP: $\varphi^2 + \theta\varphi^2$

Möbius Function:

	{c,d}	{a,b,c,d}
{c,d}	1	-1
{a,b,c,d}	0	1

Matroid: 4-3
Basis: {a}, {b}, {c}
Rank: 1   Permutations: 4   Augmented: N   Extended: Y
RG: $3 + 6\varphi + 4\varphi^2 + \varphi^3 + \theta + \theta\varphi$
RP: $1 + x$
$\chi : -1 + x$
DP: $\varphi + \theta\varphi^3$
Möbius Function:

	{d}	{a,b,c,d}
{d}	1	-1
{a,b,c,d}	0	1

Matroid: 4-4
Basis: {a}, {b}, {c}, {d}
Rank: 1   Permutations: 1   Augmented: N   Extended: N
RG: $4 + 6\varphi + 4\varphi^2 + \varphi^3 + \theta$
RP: $1 + x$
$\chi : -1 + x$
DP: $\theta\varphi^4$
Möbius Function:

	{}	{a,b,c,d}
{}	1	-1
{a,b,c,d}	0	1

Matroid: 4-5
Basis: {a,b}
Rank: 2   Permutations: 6   Augmented: Y   Extended: Y
RG: $1 + 2\varphi + \varphi^2 + 2\theta + 4\theta\varphi + 2\theta\varphi^2 + \theta^2 + 2\theta^2\varphi + \theta^2\varphi^2$
RP: $1 + 2x + x^2$
$\chi : 1 - 2x + x^2$
DP: $\varphi^2 + \theta\varphi^2$
Möbius Function:

	{c,d}	{a,c,d}	{b,c,d}	{a,b,c,d}
{c,d}	1	-1	-1	1
{a,c,d}	0	1	0	-1
{b,c,d}	0	0	1	-1
{a,b,c,d}	0	0	0	1

Matroid: 4-6
Basis: {a,b}, {a,c}
Rank: 2   Permutations: 12   Augmented: Y   Extended: Y
RG: $2 + 3\varphi + \varphi^2 + 3\theta + 4\theta\varphi + \theta\varphi^2 + \theta^2 + \theta^2\varphi$
RP: $1 + 2x + x^2$
$\chi : 1 - 2x + x^2$
DP: $\varphi + \theta\varphi^2 + \theta^2\varphi$
Möbius Function:

	{d}	{a,d}	{b,c,d}	{a,b,c,d}
{d}	1	-1	-1	1
{a,d}	0	1	0	-1
{b,c,d}	0	0	1	-1
{a,b,c,d}	0	0	0	1

Matroid: 4-7
Basis: {a,b}, {a,c}, {a,d}
Rank: 2   Permutations: 4   Augmented: Y   Extended: N
RG: $3 + 3\varphi + \varphi^2 + 4\theta + 3\theta\varphi + \theta\varphi^2 + \theta^2$
RP: $1 + 2x + x^2$
$\chi : 1 - 2x + x^2$
DP: $\theta\varphi^3 + \theta^3\varphi$
Möbius Function:

	{}	{a}	{b,c,d}	{a,b,c,d}
{}	1	-1	-1	1
{a}	0	1	0	-1
{b,c,d}	0	0	1	-1
{a,b,c,d}	0	0	0	1

Matroid: 4-8
Basis: {a,b}, {a,c}, {a,d}, {b,c}, {b,d}
Rank: 2   Permutations: 6   Augmented: N   Extended: N
RG: $5 + 4\varphi + \varphi^2 + 4\theta + \theta\varphi + \theta^2$
RP: $1 + 3x + x^2$
$\chi : 2 - 3x + x^2$
DP: $\theta^2\varphi^2 + \theta^3\varphi^2$

Möbius Function:

	{}	{a}	{b}	{c,d}	{a,b,c,d}
{}	1	-1	-1	-1	2
{a}	0	1	0	0	-1
{b}	0	0	1	0	-1
{c,d}	0	0	0	1	-1
{a,b,c,d}	0	0	0	0	1

Matroid: 4-9

Basis: {a,b}, {a,c}, {a,d}, {b,c}, {b,d}, {c,d}

Rank: 2　　　　Permutations: 1　　　　Augmented: N　　　　Extended: N

RG: $6 + 4\varphi + \varphi^2 + 4\theta + \theta^2$

RP: $1 + 4x + x^2$

$\chi : 3 - 4x + x^2$

DP: $\theta^3 \varphi^4$

Möbius Function:

	{}	{a}	{b}	{c}	{d}	{a,b,c,d}
{}	1	-1	-1	-1	-1	3
{a}	0	1	0	0	0	-1
{b}	0	0	1	0	0	-1
{c}	0	0	0	1	0	-1
{d}	0	0	0	0	1	-1
{a,b,c,d}	0	0	0	0	0	1

Matroid: 4-10

Basis: {a,b}, {a,c}, {b,c}

Rank: 2　　　　Permutations: 4　　　　Augmented: N　　　　Extended: Y

RG: $3 + 4\varphi + \varphi^2 + 3\theta + 3\theta\varphi + \theta^2 + \theta^2\varphi$

RP: $1 + 3x + x^2$

$\chi : 2 - 3x + x^2$

DP: $\varphi + \theta^2\varphi^3$

Möbius Function:

	{d}	{a,d}	{b,d}	{c,d}	{a,b,c,d}
{d}	1	-1	-1	-1	2
{a,d}	0	1	0	0	-1
{b,d}	0	0	1	0	-1
{c,d}	0	0	0	1	-1
{a,b,c,d}	0	0	0	0	1

Matroid: 4-11

Basis: {a,b}, {a,c}, {b,d}, {c,d}

Rank: 2   Permutations: 3   Augmented: N   Extended: N

RG: $4 + 4\varphi + \varphi^2 + 4\theta + 2\theta\varphi + \theta^2$

RP: $1 + 2x + x^2$

$\chi : 1 - 2x + x^2$

DP: $2\theta^2\varphi^2$

Möbius Function:

	{}	{a,d}	{b,c}	{a,b,c,d}
{}	1	-1	-1	1
{a,d}	0	1	0	-1
{b,c}	0	0	1	-1
{a,b,c,d}	0	0	0	1

Matroid: 4-12

Basis:

Rank: 0   Permutations: 1   Augmented: N   Extended: Y

RG: $1 + 4\varphi + 6\varphi^2 + 4\varphi^3 + \varphi^4$

RP: 1

$\chi : 1$

DP: φ^4

Möbius Function:

	{a,b,c,d}
{a,b,c,d}	1

Matroid: 4-13

Basis: {a,b,c}

Rank: 3   Permutations: 4   Augmented: Y   Extended: Y

RG: $1 + \varphi + 3\theta + 3\theta\varphi + 3\theta^2 + 3\theta^2\varphi + \theta^3 + \theta^3\varphi$

RP: $1 + 3x + 3x^2 + x^3$

$\chi : -1 + 3x - 3x^2 + x^3$

DP: $\varphi + \theta\varphi^3$

Möbius Function:

	{d}	{a,d}	{b,d}	{c,d}	{a,b,d}	{a,c,d}	{b,c,d}	{a,b,c,d}
{d}	1	-1	-1	-1	1	1	1	-1
{a,d}	0	1	0	0	-1	-1	0	1
{b,d}	0	0	1	0	-1	0	-1	1
{c,d}	0	0	0	1	0	-1	-1	1
{a,b,d}	0	0	0	0	1	0	0	-1
{a,c,d}	0	0	0	0	0	1	0	-1
{b,c,d}	0	0	0	0	0	0	1	-1
{a,b,c,d}	0	0	0	0	0	0	0	1

Matroid: 4-14

Basis: {a,b,c}, {a,b,d}

Rank: 3   Permutations: 6   Augmented: Y   Extended: N

RG: $2 + \varphi + 5\theta + 2\theta\varphi + 4\theta^2 + \theta^2\varphi + \theta^3$

RP: $1 + 3x + 3x^2 + x^3$

χ : $-1 + 3x - 3x^2 + x^3$

DP: $\theta\varphi^2 + \theta^2\varphi^2$

Möbius Function:

	{}	{a}	{b}	{a,b}	{c,d}	{a,c,d}	{b,c,d}	{a,b,c,d}
{}	1	-1	-1	1	-1	1	1	-1
{a}	0	1	0	-1	0	-1	0	1
{b}	0	0	1	-1	0	0	-1	1
{a,b}	0	0	0	1	0	0	0	-1
{c,d}	0	0	0	0	1	-1	-1	1
{a,c,d}	0	0	0	0	0	1	0	-1
{b,c,d}	0	0	0	0	0	0	1	-1
{a,b,c,d}	0	0	0	0	0	0	0	1

Matroid: 4-15

Basis: {a,b,c}, {a,b,d}, {a,c,d}

Rank: 3   Permutations: 4   Augmented: Y   Extended: N

RG: $3 + \varphi + 6\theta + \theta\varphi + 4\theta^2 + \theta^3$

RP: $1 + 4x + 4x^2 + x^3$

χ : $-2 + 5x - 4x^2 + x^3$

DP: $\theta^2\varphi^3 + \theta^3\varphi$

Möbius Function:

	{}	{a}	{b}	{c}	{d}	{a,b}	{a,c}	{a,d}	{b,c,d}	{a,b,c,d}
{}	1	-1	-1	-1	-1	1	1	1	2	-2
{a}	0	1	0	0	0	-1	-1	-1	0	2
{b}	0	0	1	0	0	-1	0	0	-1	1
{c}	0	0	0	1	0	0	-1	0	-1	1
{d}	0	0	0	0	1	0	0	-1	-1	1
{a,b}	0	0	0	0	0	1	0	0	0	-1
{a,c}	0	0	0	0	0	0	1	0	0	-1
{a,d}	0	0	0	0	0	0	0	1	0	-1
{b,c,d}	0	0	0	0	0	0	0	0	1	-1
{a,b,c,d}	0	0	0	0	0	0	0	0	0	1

Matroid: 4-16

Basis: $\{a,b,c\}$, $\{a,b,d\}$, $\{a,c,d\}$, $\{b,c,d\}$

Rank: 3   Permutations: 1   Augmented: N   Extended: N

RG: $4 + \varphi + 6\theta + 4\theta^2 + \theta^3$

RP: $1 + 6x + 4x^2 + x^3$

χ : $-3 + 6x - 4x^2 + x^3$

DP: $\theta^3 \varphi^4$

Möbius Function:

	{}	{a}	{b}	{c}	{d}	{a,b}	{a,c}	{a,d}	{b,c}	{b,d}	{c,d}	{a,b,c,d}
{}	1	-1	-1	-1	-1	1	1	1	1	1	1	-3
{a}	0	1	0	0	0	-1	-1	-1	0	0	0	2
{b}	0	0	1	0	0	-1	0	0	-1	-1	0	2
{c}	0	0	0	1	0	0	-1	0	-1	0	-1	2
{d}	0	0	0	0	1	0	0	-1	0	-1	-1	2
{a,b}	0	0	0	0	0	1	0	0	0	0	0	-1
{a,c}	0	0	0	0	0	0	1	0	0	0	0	-1
{a,d}	0	0	0	0	0	0	0	1	0	0	0	-1
{b,c}	0	0	0	0	0	0	0	0	1	0	0	-1
{b,d}	0	0	0	0	0	0	0	0	0	1	0	-1
{c,d}	0	0	0	0	0	0	0	0	0	0	1	-1
{a,b,c,d}	0	0	0	0	0	0	0	0	0	0	0	1

Matroid: 4-17

Basis: $\{a,b,c,d\}$

Rank: 4   Permutations: 1   Augmented: Y   Extended: N

RG: $1 + 4\theta + 6\theta^2 + 4\theta^3 + \theta^4$

RP: $1 + 4x + 6x^2 + 4x^3 + x^4$

χ : $1 - 4x + 6x^2 - 4x^3 + x^4$

DP: $\theta \varphi^4$

Möbius Function:

	{}	{a}	{b}	{c}	{d}	{a,b}	{a,c}	{a,d}	{b,c}	{b,d}	{c,d}	{a,b,c}	{a,b,d}	{a,c,d}	{b,c,d}	{a,b,c,d}
{}	1	-1	-1	-1	-1	1	1	1	1	1	1	-1	-1	-1	-1	1
{a}	0	1	0	0	0	-1	-1	-1	0	0	0	1	1	1	0	-1
{b}	0	0	1	0	0	-1	0	0	-1	-1	0	1	1	0	1	-1
{c}	0	0	0	1	0	0	-1	0	-1	0	-1	1	0	1	1	-1
{d}	0	0	0	0	1	0	0	-1	0	-1	-1	0	1	1	1	-1
{a,b}	0	0	0	0	0	1	0	0	0	0	0	-1	-1	0	0	1
{a,c}	0	0	0	0	0	0	1	0	0	0	0	-1	0	-1	0	1
{a,d}	0	0	0	0	0	0	0	1	0	0	0	0	-1	-1	0	1
{b,c}	0	0	0	0	0	0	0	0	1	0	0	-1	0	0	-1	1
{b,d}	0	0	0	0	0	0	0	0	0	1	0	0	-1	0	-1	1
{c,d}	0	0	0	0	0	0	0	0	0	0	1	0	0	-1	-1	1
{a,b,c}	0	0	0	0	0	0	0	0	0	0	0	1	0	0	0	-1
{a,b,d}	0	0	0	0	0	0	0	0	0	0	0	0	1	0	0	-1
{a,c,d}	0	0	0	0	0	0	0	0	0	0	0	0	0	1	0	-1
{b,c,d}	0	0	0	0	0	0	0	0	0	0	0	0	0	0	1	-1
{a,b,c,d}	0	0	0	0	0	0	0	0	0	0	0	0	0	0	0	1

Matroids on Five Elements

Matroid: 5-1

Basis: $\{a\}$

Rank: 1   Permutations: 5   Augmented: Y   Extended: Y

RG: $1 + 4\varphi + 6\varphi^2 + 4\varphi^3 + \varphi^4 + \theta + 4\theta\varphi + 6\theta\varphi^2 + 4\theta\varphi^3 + \theta\varphi^4$

RP: $1 + x$

$\chi : -1 + x$

DP: $\varphi^4 + \theta\varphi$

Möbius Function:

	$\{b,c,d,e\}$	$\{a,b,c,d,e\}$
$\{b,c,d,e\}$	1	-1
$\{a,b,c,d,e\}$	0	1

Matroid: 5-2

Basis: $\{a\}$, $\{b\}$

Rank: 1   Permutations: 10   Augmented: N   Extended: Y

RG: $2 + 7\varphi + 9\varphi^2 + 5\varphi^3 + \varphi^4 + \theta + 3\theta\varphi + 3\theta\varphi^2 + \theta\varphi^3$

RP: $1 + x$

$\chi : -1 + x$

DP: $\varphi^3 + \theta\varphi^2$

Möbius Function:

	$\{c,d,e\}$	$\{a,b,c,d,e\}$
$\{c,d,e\}$	1	-1
$\{a,b,c,d,e\}$	0	1

Matroid: 5-3

Basis: $\{a\}$, $\{b\}$, $\{c\}$

Rank: 1   Permutations: 10   Augmented: N   Extended: Y

RG: $3 + 9\varphi + 10\varphi^2 + 5\varphi^3 + \varphi^4 + \theta + 2\theta\varphi + \theta\varphi^2$

RP: $1 + x$

$\chi : -1 + x$

DP: $\varphi^2 + \theta\varphi^3$

Möbius Function:

	$\{d,e\}$	$\{a,b,c,d,e\}$
$\{d,e\}$	1	-1
$\{a,b,c,d,e\}$	0	1

Matroid: 5-4

Basis: {a}, {b}, {c}, {d}

Rank: 1   Permutations: 5   Augmented: N   Extended: Y

RG: $4 + 10\varphi + 10\varphi^2 + 5\varphi^3 + \varphi^4 + \theta + \theta\varphi$

RP: $1 + x$

$\chi : -1 + x$

DP: $\varphi + \theta\varphi^4$

Möbius Function:

	{e}	{a,b,c,d,e}
{e}	1	-1
{a,b,c,d,e}	0	1

Matroid: 5-5

Basis: {a}, {b}, {c}, {d}, {e}

Rank: 1   Permutations: 1   Augmented: N   Extended: N

RG: $5 + 10\varphi + 10\varphi^2 + 5\varphi^3 + \varphi^4 + \theta$

RP: $1 + x$

$\chi : -1 + x$

DP: $\theta\varphi^5$

Möbius Function:

	{}	{a,b,c,d,e}
{}	1	-1
{a,b,c,d,e}	0	1

Matroid: 5-6

Basis: {a,b}

Rank: 2   Permutations: 10   Augmented: Y   Extended: Y

RG: $1 + 3\varphi + 3\varphi^2 + \varphi^3 + 2\theta + 6\theta\varphi + 6\theta\varphi^2 + 2\theta\varphi^3 + \theta^2 + 3\theta^2\varphi + 3\theta^2\varphi^2 + \theta^2\varphi^3$

RP: $1 + 2x + x^2$

$\chi : 1 - 2x + x^2$

DP: $\varphi^3 + \theta\varphi^2$

Möbius Function:

	{c,d,e}	{a,c,d,e}	{b,c,d,e}	{a,b,c,d,e}
{c,d,e}	1	-1	-1	1
{a,c,d,e}	0	1	0	-1
{b,c,d,e}	0	0	1	-1
{a,b,c,d,e}	0	0	0	1

Matroid: 5-7

Basis: {a,b}, {a,c}

Rank: 2   Permutations: 30   Augmented: Y   Extended: Y

RG: $2 + 5\varphi + 4\varphi^2 + \varphi^3 + 3\theta + 7\theta\varphi + 5\theta\varphi^2 + \theta\varphi^3 + \theta^2 + 2\theta^2\varphi + \theta^2\varphi^2$

RP: $1 + 2x + x^2$

$\chi: 1 - 2x + x^2$

DP: $\varphi^2 + \theta\varphi^2 + \theta^2\varphi$

Möbius Function:

	{d,e}	{a,d,e}	{b,c,d,e}	{a,b,c,d,e}
{d,e}	1	-1	-1	1
{a,d,e}	0	1	0	-1
{b,c,d,e}	0	0	1	-1
{a,b,c,d,e}	0	0	0	1

Matroid: 5-8

Basis: {a,b}, {a,c}, {a,d}

Rank: 2   Permutations: 20   Augmented: Y   Extended: Y

RG: $3 + 6\varphi + 4\varphi^2 + \varphi^3 + 4\theta + 7\theta\varphi + 4\theta\varphi^2 + \theta\varphi^3 + \theta^2 + \theta^2\varphi$

RP: $1 + 2x + x^2$

$\chi: 1 - 2x + x^2$

DP: $\varphi + \theta\varphi^3 + \theta^3\varphi$

Möbius Function:

	{e}	{a,e}	{b,c,d,e}	{a,b,c,d,e}
{e}	1	-1	-1	1
{a,e}	0	1	0	-1
{b,c,d,e}	0	0	1	-1
{a,b,c,d,e}	0	0	0	1

Matroid: 5-9

Basis: {a,b}, {a,c}, {a,d}, {a,e}

Rank: 2   Permutations: 5   Augmented: Y   Extended: N

RG: $4 + 6\varphi + 4\varphi^2 + \varphi^3 + 5\theta + 6\theta\varphi + 4\theta\varphi^2 + \theta\varphi^3 + \theta^2$

RP: $1 + 2x + x^2$

$\chi: 1 - 2x + x^2$

DP: $\theta\varphi^4 + \theta^4\varphi$

Möbius Function:

	{}	{a}	{b,c,d,e}	{a,b,c,d,e}
{}	1	-1	-1	1
{a}	0	1	0	-1
{b,c,d,e}	0	0	1	-1
{a,b,c,d,e}	0	0	0	1

Matroid: 5-10

Basis: {a,b}, {a,c}, {a,d}, {a,e}, {b,c}, {b,d}, {b,e}

Rank: 2   Permutations: 10   Augmented: N   Extended: N

RG: $7 + 9\varphi + 5\varphi^2 + \varphi^3 + 5\theta + 3\theta\varphi + \theta\varphi^2 + \theta^2$

RP: $1 + 3x + x^2$

$\chi : 2 - 3x + x^2$

DP: $\theta^2\varphi^3 + \theta^4\varphi^2$

Möbius Function:

	{}	{a}	{b}	{c,d,e}	{a,b,c,d,e}
{}	1	-1	-1	-1	2
{a}	0	1	0	0	-1
{b}	0	0	1	0	-1
{c,d,e}	0	0	0	1	1
{a,b,c,d,e}	0	0	0	0	1

Matroid: 5-11

Basis: {a,b}, {a,c}, {a,d}, {a,e}, {b,c}, {b,d}, {b,e}, {c,d}, {c,e}

Rank: 2   Permutations: 10   Augmented: N   Extended: N

RG: $9 + 10\varphi + 5\varphi^2 + \varphi^3 + 5\theta + \theta\varphi + \theta^2$

RP: $1 + 4x + x^2$

$\chi : 3 - 4x + x^2$

DP: $\theta^3\varphi^2 + \theta^4\varphi^3$

Möbius Function:

	{}	{a}	{b}	{c}	{d,e}	{a,b,c,d,e}
{}	1	-1	-1	-1	-1	3
{a}	0	1	0	0	0	-1
{b}	0	0	1	0	0	-1
{c}	0	0	0	1	0	-1
{d,e}	0	0	0	0	1	-1
{a,b,c,d,e}	0	0	0	0	0	1

Matroid: 5-12

Basis: {a,b}, {a,c}, {a,d}, {a,e}, {b,c}, {b,d}, {b,e}, {c,d}, {c,e}, {d,e}

Rank: 2   Permutations: 1   Augmented: N   Extended: N

RG: $10 + 10\varphi + 5\varphi^2 + \varphi^3 + 5\theta + \theta^2$

RP: $1 + 5x + x^2$

$\chi : 4 - 5x + x^2$

DP: $\theta^4\varphi^5$

Möbius Function:

	{}	{a}	{b}	{c}	{d}	{e}	{a,b,c,d,e}
{}	1	-1	-1	-1	-1	-1	4
{a}	0	1	0	0	0	0	-1
{b}	0	0	1	0	0	0	-1
{c}	0	0	0	1	0	0	-1
{d}	0	0	0	0	1	0	-1
{e}	0	0	0	0	0	1	-1
{a,b,c,d,e}	0	0	0	0	0	0	1

Matroid: 5-13

Basis: {a,b}, {a,c}, {a,d}, {a,e}, {b,c}, {b,d}, {c,e}, {d,e}

Rank: 2   Permutations: 15   Augmented: N   Extended: N

RG: $8 + 10\varphi + 5\varphi^2 + \varphi^3 + 5\theta + 2\theta\varphi + \theta^2$

RP: $1 + 3x + x^2$

$\chi : 2 - 3x + x^2$

DP: $2\theta^3\varphi^2 + \theta^4\varphi$

Möbius Function:

	{}	{a}	{b,e}	{c,d}	{a,b,c,d,e}
{}	1	-1	-1	-1	2
{a}	0	1	0	0	-1
{b,e}	0	0	1	0	-1
{c,d}	0	0	0	1	-1
{a,b,c,d,e}	0	0	0	0	1

Matroid: 5-14

Basis: {a,b}, {a,c}, {a,d}, {b,c}, {b,d}

Rank: 2   Permutations: 30   Augmented: N   Extended: Y

RG: $5 + 9\varphi + 5\varphi^2 + \varphi^3 + 4\theta + 5\theta\varphi + \theta\varphi^2 + \theta^2 + \theta^2\varphi$

RP: $1 + 3x + x^2$

$\chi : 2 - 3x + x^2$

DP: $\varphi + \theta^2\varphi^2 + \theta^3\varphi^2$

Möbius Function:

	{e}	{a,e}	{b,e}	{c,d,e}	{a,b,c,d,e}
{e}	1	-1	-1	-1	2
{a,e}	0	1	0	0	-1
{b,e}	0	0	1	0	-1
{c,d,e}	0	0	0	1	-1
{a,b,c,d,e}	0	0	0	0	1

Matroid: 5-15

Basis: {a,b}, {a,c}, {a,d}, {b,c}, {b,d}, {c,d}

Rank: 2   Permutations: 5   Augmented: N   Extended: Y

RG: $6 + 10\varphi + 5\varphi^2 + \varphi^3 + 4\theta + 4\theta\varphi + \theta^2 + \theta^2\varphi$

RP: $1 + 4x + x^2$

$\chi: 3 - 4x + x^2$

DP: $\varphi + \theta^3\varphi^4$

Möbius Function:

	{e}	{a,e}	{b,e}	{c,e}	{d,e}	{a,b,c,d,e}
{e}	1	-1	-1	-1	-1	3
{a,e}	0	1	0	0	0	-1
{b,e}	0	0	1	0	0	-1
{c,e}	0	0	0	1	0	-1
{d,e}	0	0	0	0	1	-1
{a,b,c,d,e}	0	0	0	0	0	1

Matroid: 5-16

Basis: {a,b}, {a,c}, {a,d}, {b,e}, {c,e}, {d,e}

Rank: 2   Permutations: 10   Augmented: N   Extended: N

RG: $6 + 9\varphi + 5\varphi^2 + \varphi^3 + 5\theta + 4\theta\varphi + \theta\varphi^2 + \theta^2$

RP: $1 + 2x + x^2$

$\chi: 1 - 2x + x^2$

DP: $\theta^2\varphi^3 + \theta^3\varphi^2$

Möbius Function:

	{}	{a,e}	{b,c,d}	{a,b,c,d,e}
{}	1	-1	-1	1
{a,e}	0	1	0	-1
{b,c,d}	0	0	1	-1
{a,b,c,d,e}	0	0	0	1

Matroid: 5-17
Basis: {a,b}, {a,c}, {b,c}
Rank: 2   Permutations: 10   Augmented: N   Extended: Y
RG: $3 + 7\varphi + 5\varphi^2 + \varphi^3 + 3\theta + 6\theta\varphi + 3\theta\varphi^2 + \theta^2 + 2\theta^2\varphi + \theta^2\varphi^2$
RP: $1 + 3x + x^2$
$\chi : 2 - 3x + x^2$
DP: $\varphi^2 + \theta^2\varphi^3$
Möbius Function:

	{d,e}	{a,d,e}	{b,d,e}	{c,d,e}	{a,b,c,d,e}
{d,e}	1	-1	-1	-1	2
{a,d,e}	0	1	0	0	-1
{b,d,e}	0	0	1	0	-1
{c,d,e}	0	0	0	1	-1
{a,b,c,d,e}	0	0	0	0	1

Matroid: 5-18
Basis: {a,b}, {a,c}, {b,d}, {c,d}
Rank: 2   Permutations: 15   Augmented: N   Extended: Y
RG: $4 + 8\varphi + 5\varphi^2 + \varphi^3 + 4\theta + 6\theta\varphi + 2\theta\varphi^2 + \theta^2 + \theta^2\varphi$
RP: $1 + 2x + x^2$
$\chi : 1 - 2x + x^2$
DP: $\varphi + 2\theta^2\varphi^2$
Möbius Function:

	{e}	{a,d,e}	{b,c,e}	{a,b,c,d,e}
{e}	1	-1	-1	1
{a,d,e}	0	1	0	-1
{b,c,e}	0	0	1	-1
{a,b,c,d,e}	0	0	0	1

Matroid: 5-19
Basis: {a,b,c}
Rank: 3   Permutations: 10   Augmented: Y   Extended: Y
RG: $1 + 2\varphi + \varphi^2 + 3\theta + 6\theta\varphi + 3\theta\varphi^2 + 3\theta^2 + 6\theta^2\varphi + 3\theta^2\varphi^2 + \theta^3 + 2\theta^3\varphi + \theta^3\varphi^2$
RP: $1 + 3x + 3x^2 + x^3$
$\chi : -1 + 3x - 3x^2 + x^3$
DP: $\varphi^2 + \theta\varphi^3$

Möbius Function:

	{d,e}	{a,d,e}	{b,d,e}	{c,d,e}	{a,b,d,e}	{a,c,d,e}	{b,c,d,e}	{a,b,c,d,e}
{d,e}	1	-1	-1	-1	1	1	1	-1
{a,d,e}	0	1	0	0	-1	-1	0	1
{b,d,e}	0	0	1	0	-1	0	-1	1
{c,d,e}	0	0	0	1	0	-1	-1	1
{a,b,d,e}	0	0	0	0	1	0	0	-1
{a,c,d,e}	0	0	0	0	0	1	0	-1
{b,c,d,e}	0	0	0	0	0	0	1	-1
{a,b,c,d,e}	0	0	0	0	0	0	0	1

Matroid: 5-20

Basis: {a,b,c}, {a,b,d}

Rank: 3   Permutations: 30   Augmented: Y   Extended: Y

RG: $2 + 3\varphi + \varphi^2 + 5\theta + 7\theta\varphi + 2\theta\varphi^2 + 4\theta^2 + 5\theta^2\varphi + \theta^2\varphi^2 + \theta^3 + \theta^3\varphi$

RP: $1 + 3x + 3x^2 + x^3$

$\chi : -1 + 3x - 3x^2 + x^3$

DP: $\varphi + \theta\varphi^2 + \theta^2\varphi^2$

Möbius Function:

	{e}	{a,e}	{b,e}	{a,b,e}	{c,d,e}	{a,c,d,e}	{b,c,d,e}	{a,b,c,d,e}
{e}	1	-1	-1	1	-1	1	1	-1
{a,e}	0	1	0	-1	0	-1	0	1
{b,e}	0	0	1	-1	0	0	-1	1
{a,b,e}	0	0	0	1	0	0	0	-1
{c,d,e}	0	0	0	0	1	-1	-1	1
{a,c,d,e}	0	0	0	0	0	1	0	-1
{b,c,d,e}	0	0	0	0	0	0	1	-1
{a,b,c,d,e}	0	0	0	0	0	0	0	1

Matroid: 5-21

Basis: {a,b,c}, {a,b,d}, {a,b,e}

Rank: 3   Permutations: 10   Augmented: Y   Extended: N

RG: $3 + 3\varphi + \varphi^2 + 7\theta + 6\theta\varphi + 2\theta\varphi^2 + 5\theta^2 + 3\theta^2\varphi + \theta^2\varphi^2 + \theta^3$

RP: $1 + 3x + 3x^2 + x^3$

$\chi : -1 + 3x - 3x^2 + x^3$

DP: $\theta\varphi^3 + \theta^3\varphi^2$

Möbius Function:

	{}	{a}	{b}	{a,b}	{c,d,e}	{a,c,d,e}	{b,c,d,e}	{a,b,c,d,e}
{}	1	-1	-1	1	-1	1	1	-1
{a}	0	1	0	-1	0	-1	0	1
{b}	0	0	1	-1	0	0	-1	1
{a,b}	0	0	0	1	0	0	0	-1
{c,d,e}	0	0	0	0	1	-1	-1	1
{a,c,d,e}	0	0	0	0	0	1	0	-1
{b,c,d,e}	0	0	0	0	0	0	1	-1
{a,b,c,d,e}	0	0	0	0	0	0	0	1

Matroid: 5-22
Basis: {a,b,c}, {a,b,d}, {a,b,e}, {a,c,d}, {a,c,e}
Rank: 3   Permutations: 30   Augmented: Y   Extended: N
RG: $5 + 4\varphi + \varphi^2 + 9\theta + 5\theta\varphi + \theta\varphi^2 + 5\theta^2 + \theta^2\varphi + \theta^3$
RP: $1 + 4x + 4x^2 + x^3$
$\chi : -2 + 5x - 4x^2 + x^3$
DP: $\theta^2\varphi^2 + \theta^3\varphi^2 + \theta^5\varphi$
Möbius Function:

	{}	{a}	{b}	{c}	{a,b}	{a,c}	{d,e}	{a,d,e}	{b,c,d,e}	{a,b,c,d,e}
{}	1	-1	-1	-1	1	1	-1	1	2	-2
{a}	0	1	0	0	-1	-1	0	-1	0	2
{b}	0	0	1	0	-1	0	0	0	-1	1
{c}	0	0	0	1	0	-1	0	0	-1	1
{a,b}	0	0	0	0	1	0	0	0	0	-1
{a,c}	0	0	0	0	0	1	0	0	0	-1
{d,e}	0	0	0	0	0	0	1	-1	-1	1
{a,d,e}	0	0	0	0	0	0	0	1	0	-1
{b,c,d,e}	0	0	0	0	0	0	0	0	1	-1
{a,b,c,d,e}	0	0	0	0	0	0	0	0	0	1

Matroid: 5-23
Basis: {a,b,c}, {a,b,d}, {a,b,e}, {a,c,d}, {a,c,e}, {a,d,e}
Rank: 3   Permutations: 5   Augmented: Y   Extended: N
RG: $6 + 4\varphi + \varphi^2 + 10\theta + 4\theta\varphi + \theta\varphi^2 + 5\theta^2 + \theta^3$
RP: $1 + 5x + 5x^2 + x^3$
$\chi : -3 + 7x - 5x^2 + x^3$
DP: $\theta^3\varphi^4 + \theta^6\varphi$
Möbius Function:

	{}	{a}	{b}	{c}	{d}	{e}	{a,b}	{a,c}	{a,d}	{a,e}	{b,c,d,e}	{a,b,c,d,e}
{}	1	-1	-1	-1	-1	-1	1	1	1	1	3	-3
{a}	0	1	0	0	0	0	-1	-1	-1	-1	0	3
{b}	0	0	1	0	0	0	-1	0	0	0	-1	1
{c}	0	0	0	1	0	0	0	-1	0	0	-1	1
{d}	0	0	0	0	1	0	0	0	-1	0	-1	1
{e}	0	0	0	0	0	1	0	0	0	-1	-1	1
{a,b}	0	0	0	0	0	0	1	0	0	0	0	-1
{a,c}	0	0	0	0	0	0	0	1	0	0	0	-1
{a,d}	0	0	0	0	0	0	0	0	1	0	0	-1
{a,e}	0	0	0	0	0	0	0	0	0	1	0	-1
{b,c,d,e}	0	0	0	0	0	0	0	0	0	0	1	-1
{a,b,c,d,e}	0	0	0	0	0	0	0	0	0	0	0	1

Matroid: 5-24

Basis: {a,b,c}, {a,b,d}, {a,b,e}, {a,c,d}, {a,c,e}, {a,d,e}, {b,c,d}, {b,c,e}, {b,d,e}

Rank: 3   Permutations: 10   Augmented: N   Extended: N

RG: $9 + 5\varphi + \varphi^2 + 10\theta + \theta\varphi + 5\theta^2 + \theta^3$

RP: $1 + 8x + 5x^2 + x^3$

$\chi: -5 + 9x - 5x^2 + x^3$

DP: $\theta^5\varphi^3 + \theta^6\varphi^2$

Möbius Function:

	{}	{a}	{b}	{c}	{d}	{e}	{a,b}	{a,c}	{a,d}	{a,e}	{b,c}	{b,d}	{b,e}	{c,d,e}	{a,b,c,d,e}
{}	1	-1	-1	-1	-1	-1	1	1	1	1	1	1	1	2	-5
{a}	0	1	0	0	0	0	-1	-1	-1	-1	0	0	0	0	3
{b}	0	0	1	0	0	0	-1	0	0	0	-1	-1	-1	0	3
{c}	0	0	0	1	0	0	0	-1	0	0	-1	0	0	-1	2
{d}	0	0	0	0	1	0	0	0	-1	0	0	-1	0	-1	2
{e}	0	0	0	0	0	1	0	0	0	-1	0	0	-1	-1	2
{a,b}	0	0	0	0	0	0	1	0	0	0	0	0	0	0	-1
{a,c}	0	0	0	0	0	0	0	1	0	0	0	0	0	0	-1
{a,d}	0	0	0	0	0	0	0	0	1	0	0	0	0	0	-1
{a,e}	0	0	0	0	0	0	0	0	0	1	0	0	0	0	-1
{b,c}	0	0	0	0	0	0	0	0	0	0	1	0	0	0	-1
{b,d}	0	0	0	0	0	0	0	0	0	0	0	1	0	0	-1
{b,e}	0	0	0	0	0	0	0	0	0	0	0	0	1	0	-1
{c,d,e}	0	0	0	0	0	0	0	0	0	0	0	0	0	1	-1
{a,b,c,d,e}	0	0	0	0	0	0	0	0	0	0	0	0	0	0	1

Matroid: 5-25

Basis: {a,b,c}, {a,b,d}, {a,b,e}, {a,c,d}, {a,c,e}, {a,d,e}, {b,c,d}, {b,c,e}, {b,d,e}, {c,d,e}

Rank: 3   Permutations: 1   Augmented: N   Extended: N

RG: $10 + 5\varphi + \varphi^2 + 10\theta + 5\theta^2 + \theta^3$

RP: $1 + 10x + 5x^2 + x^3$

$\chi: -6 + 10x - 5x^2 + x^3$

DP: $\theta^6\varphi^5$

Matroid: 5-26

Basis: {a,b,c}, {a,b,d}, {a,b,e}, {a,c,d}, {a,c,e}, {b,c,d}, {b,c,e}

Rank: 3   Permutations: 10   Augmented: N   Extended: N

RG: $7 + 5\varphi + \varphi^2 + 9\theta + 3\theta\varphi + 5\theta^2 + \theta^2\varphi + \theta^3$

RP: $1 + 6x + 4x^2 + x^3$

$\chi: -3 + 6x - 4x^2 + x^3$

DP: $\theta^3\varphi^2 + \theta^5\varphi^3$

Matroid: 5-27

Basis: $\{a,b,c\}$, $\{a,b,d\}$, $\{a,b,e\}$, $\{a,c,d\}$, $\{a,c,e\}$, $\{b,c,d\}$, $\{b,d,e\}$, $\{c,d,e\}$

Rank: 3   Permutations: 15   Augmented: N   Extended: N

RG: $8 + 5\varphi + \varphi^2 + 10\theta + 2\theta\varphi + 5\theta^2 + \theta^3$

RP: $1 + 6x + 5x^2 + x^3$

$\chi : -4 + 8x - 5x^2 + x^3$

DP: $\theta^4\varphi + 2\theta^5\varphi^2$

Möbius Function:

	{}	{a}	{b}	{c}	{d}	{e}	{a,b}	{a,c}	{b,d}	{c,d}	{a,d,e}	{b,c,e}	{a,b,c,d,e}
{}	1	-1	-1	-1	-1	-1	1	1	1	1	2	2	-4
{a}	0	1	0	0	0	0	-1	-1	0	0	-1	0	2
{b}	0	0	1	0	0	0	-1	0	-1	0	0	-1	2
{c}	0	0	0	1	0	0	0	-1	0	-1	0	-1	2
{d}	0	0	0	0	1	0	0	0	-1	-1	-1	0	2
{e}	0	0	0	0	0	1	0	0	0	0	-1	-1	1
{a,b}	0	0	0	0	0	0	1	0	0	0	0	0	-1
{a,c}	0	0	0	0	0	0	0	1	0	0	0	0	-1
{b,d}	0	0	0	0	0	0	0	0	1	0	0	0	-1
{c,d}	0	0	0	0	0	0	0	0	0	1	0	0	-1
{a,d,e}	0	0	0	0	0	0	0	0	0	0	1	0	-1
{b,c,e}	0	0	0	0	0	0	0	0	0	0	0	1	-1
{a,b,c,d,e}	0	0	0	0	0	0	0	0	0	0	0	0	1

Matroid: 5-28

Basis: $\{a,b,c\}$, $\{a,b,d\}$, $\{a,c,d\}$

Rank: 3   Permutations: 20   Augmented: Y   Extended: Y

RG: $3 + 4\varphi + \varphi^2 + 6\theta + 7\theta\varphi + \theta\varphi^2 + 4\theta^2 + 4\theta^2\varphi + \theta^3 + \theta^3\varphi$

RP: $1 + 4x + 4x^2 + x^3$

$\chi : -2 + 5x - 4x^2 + x^3$

DP: $\varphi + \theta^2\varphi^3 + \theta^3\varphi$

Möbius Function:

	{e}	{a,e}	{b,e}	{c,e}	{d,e}	{a,b,e}	{a,c,e}	{a,d,e}	{b,c,d,e}	{a,b,c,d,e}
{e}	1	-1	-1	-1	-1	1	1	1	2	-2
{a,e}	0	1	0	0	0	-1	-1	-1	0	2
{b,e}	0	0	1	0	0	-1	0	0	-1	1
{c,e}	0	0	0	1	0	0	-1	0	-1	1
{d,e}	0	0	0	0	1	0	0	-1	-1	1
{a,b,e}	0	0	0	0	0	1	0	0	0	-1
{a,c,e}	0	0	0	0	0	0	1	0	0	-1
{a,d,e}	0	0	0	0	0	0	0	1	0	-1
{b,c,d,e}	0	0	0	0	0	0	0	0	1	-1
{a,b,c,d,e}	0	0	0	0	0	0	0	0	0	1

Matroid: 5-29

Basis: $\{a,b,c\}$, $\{a,b,d\}$, $\{a,c,d\}$, $\{b,c,d\}$

Rank: 3   Permutations: 5   Augmented: N   Extended: Y

RG: $4 + 5\varphi + \varphi^2 + 6\theta + 6\theta\varphi + 4\theta^2 + 4\theta^2\varphi + \theta^3 + \theta^3\varphi$

RP: $1 + 6x + 4x^2 + x^3$

$\chi : -3 + 6x - 4x^2 + x^3$

DP: $\varphi + \theta^3\varphi^4$

Möbius Function:

	{e}	{a,e}	{b,e}	{c,e}	{d,e}	{a,b,e}	{a,c,e}	{a,d,e}	{b,c,e}	{b,d,e}	{c,d,e}	{a,b,c,d,e}
{e}	1	-1	-1	-1	-1	1	1	1	1	1	1	-3
{a,e}	0	1	0	0	0	-1	-1	-1	0	0	0	2
{b,e}	0	0	1	0	0	-1	0	0	-1	-1	0	2
{c,e}	0	0	0	1	0	0	-1	0	-1	0	-1	2
{d,e}	0	0	0	0	1	0	0	-1	0	-1	-1	2
{a,b,e}	0	0	0	0	0	1	0	0	0	0	0	-1
{a,c,e}	0	0	0	0	0	0	1	0	0	0	0	-1
{a,d,e}	0	0	0	0	0	0	0	1	0	0	0	-1
{b,c,e}	0	0	0	0	0	0	0	0	1	0	0	-1
{b,d,e}	0	0	0	0	0	0	0	0	0	1	0	-1
{c,d,e}	0	0	0	0	0	0	0	0	0	0	1	-1
{a,b,c,d,e}	0	0	0	0	0	0	0	0	0	0	0	1

Matroid: 5-30

Basis: $\{a,b,c\}$, $\{a,b,d\}$, $\{a,c,d\}$, $\{b,c,e\}$, $\{b,d,e\}$, $\{c,d,e\}$

Rank: 3   Permutations: 10   Augmented: N   Extended: N

RG: $6 + 5\varphi + \varphi^2 + 9\theta + 4\theta\varphi + 5\theta^2 + \theta^2\varphi + \theta^3$

RP: $1 + 4x + 4x^2 + x^3$

$\chi : -2 + 5x - 4x^2 + x^3$

DP: $\theta^3\varphi^2 + \theta^4\varphi^3$

Möbius Function:

	{}	{b}	{c}	{d}	{a,e}	{a,b,e}	{a,c,e}	{a,d,e}	{b,c,d}	{a,b,c,d,e}
{}	1	-1	-1	-1	-1	1	1	1	2	-2
{b}	0	1	0	0	0	-1	0	0	-1	1
{c}	0	0	1	0	0	0	-1	0	-1	1
{d}	0	0	0	1	0	0	0	-1	-1	1
{a,e}	0	0	0	0	1	-1	-1	-1	0	2
{a,b,e}	0	0	0	0	0	1	0	0	0	-1
{a,c,e}	0	0	0	0	0	0	1	0	0	-1
{a,d,e}	0	0	0	0	0	0	0	1	0	-1
{b,c,d}	0	0	0	0	0	0	0	0	1	-1
{a,b,c,d,e}	0	0	0	0	0	0	0	0	0	1

Matroid: 5-31

Basis: {a,b,c}, {a,b,d}, {a,c,e}, {a,d,e}

Rank: 3 Permutations: 15 Augmented: Y Extended: N

RG: $4 + 4\varphi + \varphi^2 + 8\theta + 6\theta\varphi + \theta\varphi^2 + 5\theta^2 + 2\theta^2\varphi + \theta^3$

RP: $1 + 3x + 3x^2 + x^3$

$\chi : -1 + 3x - 3x^2 + x^3$

DP: $2\theta^2\varphi^2 + \theta^4\varphi$

Möbius Function:

	{}	{a}	{b,e}	{c,d}	{a,b,e}	{a,c,d}	{b,c,d,e}	{a,b,c,d,e}
{}	1	-1	-1	-1	1	1	1	-1
{a}	0	1	0	0	-1	-1	0	1
{b,e}	0	0	1	0	-1	0	-1	1
{c,d}	0	0	0	1	0	-1	-1	1
{a,b,e}	0	0	0	0	1	0	0	-1
{a,c,d}	0	0	0	0	0	1	0	-1
{b,c,d,e}	0	0	0	0	0	0	1	-1
{a,b,c,d,e}	0	0	0	0	0	0	0	1

Matroid: 5-32

Basis:

Rank: 0 Permutations: 1 Augmented: N Extended: Y

RG: $1 + 5\varphi + 10\varphi^2 + 10\varphi^3 + 5\varphi^4 + \varphi^5$

RP: 1

$\chi : 1$

DP: φ^5

Möbius Function:

	{a,b,c,d,e}
{a,b,c,d,e}	1

Matroid: 5-33

Basis: {a,b,c,d}

Rank: 4 Permutations: 5 Augmented: Y Extended: Y

RG: $1 + \varphi + 4\theta + 4\theta\varphi + 6\theta^2 + 6\theta^2\varphi + 4\theta^3 + 4\theta^3\varphi + \theta^4 + \theta^4\varphi$

RP: $1 + 4x + 6x^2 + 4x^3 + x^4$

$\chi : 1 - 4x + 6x^2 - 4x^3 + x^4$

DP: $\varphi + \theta\varphi^4$

Matroid: 5-34
Basis: {a,b,c,d}, {a,b,c,e}
Rank: 4   Permutations: 10   Augmented: Y   Extended: N
RG: $2 + \varphi + 7\theta + 3\theta\varphi + 9\theta^2 + 3\theta^2\varphi + 5\theta^3 + \theta^3\varphi + \theta^4$
RP: $1 + 4x + 6x^2 + 4x^3 + x^4$
$\chi : 1 - 4x + 6x^2 - 4x^3 + x^4$
DP: $\theta\varphi^2 + \theta^2\varphi^3$

Matroid: 5-35
Basis: {a,b,c,d}, {a,b,c,e}, {a,b,d,e}
Rank: 4   Permutations: 10   Augmented: Y   Extended: N
RG: $3 + \varphi + 9\theta + 2\theta\varphi + 10\theta^2 + \theta^2\varphi + 5\theta^3 + \theta^4$
RP: $1 + 5x + 8x^2 + 5x^3 + x^4$
$\chi : 2 - 7x + 9x^2 - 5x^3 + x^4$
DP: $\theta^2\varphi^3 + \theta^3\varphi^2$

Matroid: 5-36
Basis: {a,b,c,d}, {a,b,c,e}, {a,b,d,e}, {a,c,d,e}
Rank: 4   Permutations: 5   Augmented: Y   Extended: N
RG: $4 + \varphi + 10\theta + \theta\varphi + 10\theta^2 + 5\theta^3 + \theta^4$
RP: $1 + 7x + 10x^2 + 5x^3 + x^4$
$\chi : 3 - 9x + 10x^2 - 5x^3 + x^4$
DP: $\theta^3\varphi^4 + \theta^4\varphi$

Matroid: 5-37
Basis: {a,b,c,d}, {a,b,c,e}, {a,b,d,e}, {a,c,d,e}, {b,c,d,e}
Rank: 4   Permutations: 1   Augmented: N   Extended: N
RG: $5 + \varphi + 10\theta + 10\theta^2 + 5\theta^3 + \theta^4$
RP: $1 + 10x + 10x^2 + 5x^3 + x^4$
$\chi : 4 - 10x + 10x^2 - 5x^3 + x^4$
DP: $\theta^4\varphi^5$

Matroid: 5-38
Basis: {a,b,c,d,e}
Rank: 5   Permutations: 1   Augmented: Y   Extended: N
RG: $1 + 5\theta + 10\theta^2 + 10\theta^3 + 5\theta^4 + \theta^5$
RP: $1 + 5x + 10x^2 + 10x^3 + 5x^4 + x^5$
$\chi : -1 + 5x - 10x^2 + 10x^3 - 5x^4 + x^5$
DP: $\theta\varphi^5$

Matroids on Six Elements

Matroid: 6-1

Basis: $\{a\}$

Rank: 1   Permutations: 6   Augmented: Y   Extended: Y

RG: $1 + 5\varphi + 10\varphi^2 + 10\varphi^3 + 5\varphi^4 + \varphi^5 + \theta + 5\theta\varphi + 10\theta\varphi^2 + 10\theta\varphi^3 + 5\theta\varphi^4 + \theta\varphi^5$

RP: $1 + x$

$\chi : -1 + x$

DP: $\varphi^5 + \theta\varphi$

Möbius Function:

	$\{b,c,d,e,f\}$	$\{a,b,c,d,e,f\}$
$\{b,c,d,e,f\}$	1	-1
$\{a,b,c,d,e,f\}$	0	1

Matroid: 6-2

Basis: $\{a\}, \{b\}$

Rank: 1   Permutations: 15   Augmented: N   Extended: Y

RG: $2 + 9\varphi + 16\varphi^2 + 14\varphi^3 + 6\varphi^4 + \varphi^5 + \theta + 4\theta\varphi + 6\theta\varphi^2 + 4\theta\varphi^3 + \theta\varphi^4$

RP: $1 + x$

$\chi : -1 + x$

DP: $\varphi^4 + \theta\varphi^2$

Möbius Function:

	$\{c,d,e,f\}$	$\{a,b,c,d,e,f\}$
$\{c,d,e,f\}$	1	-1
$\{a,b,c,d,e,f\}$	0	1

Matroid: 6-3

Basis: $\{a\}, \{b\}, \{c\}$

Rank: 1   Permutations: 20   Augmented: N   Extended: Y

RG: $3 + 12\varphi + 19\varphi^2 + 15\varphi^3 + 6\varphi^4 + \varphi^5 + \theta + 3\theta\varphi + 3\theta\varphi^2 + \theta\varphi^3$

RP: $1 + x$

$\chi : -1 + x$

DP: $\varphi^3 + \theta\varphi^3$

Möbius Function:

	$\{d,e,f\}$	$\{a,b,c,d,e,f\}$
$\{d,e,f\}$	1	-1
$\{a,b,c,d,e,f\}$	0	1

Matroid: 6-4

Basis: {a}, {b}, {c}, {d}

Rank: 1   Permutations: 15   Augmented: N   Extended: Y

RG: $4 + 14\varphi + 20\varphi^2 + 15\varphi^3 + 6\varphi^4 + \varphi^5 + \theta + 2\theta\varphi + \theta\varphi^2$

RP: $1 + x$

$\chi : -1 + x$

DP: $\varphi^2 + \theta\varphi^4$

Möbius Function:

	{e,f}	{a,b,c,d,e,f}
{e,f}	1	-1
{a,b,c,d,e,f}	0	1

Matroid: 6-5

Basis: {a}, {b}, {c}, {d}, {e}

Rank: 1   Permutations: 6   Augmented: N   Extended: Y

RG: $5 + 15\varphi + 20\varphi^2 + 15\varphi^3 + 6\varphi^4 + \varphi^5 + \theta + \theta\varphi$

RP: $1 + x$

$\chi : -1 + x$

DP: $\varphi + \theta\varphi^5$

Möbius Function:

	{f}	{a,b,c,d,e,f}
{f}	1	-1
{a,b,c,d,e,f}	0	1

Matroid: 6-6

Basis: {a}, {b}, {c}, {d}, {e}, {f}

Rank: 1   Permutations: 1   Augmented: N   Extended: N

RG: $6 + 15\varphi + 20\varphi^2 + 15\varphi^3 + 6\varphi^4 + \varphi^5 + \theta$

RP: $1 + x$

$\chi : -1 + x$

DP: $\theta\varphi^6$

Möbius Function:

	{}	{a,b,c,d,e,f}
{}	1	-1
{a,b,c,d,e,f}	0	1

Matroid: 6-7

Basis: {a,b}

Rank: 2   Permutations: 15   Augmented: Y   Extended: Y

RG: $1 + 4\varphi + 6\varphi^2 + 4\varphi^3 + \varphi^4 + 2\theta + 8\theta\varphi + 12\theta\varphi^2 + 8\theta\varphi^3 + 2\theta\varphi^4 + \theta^2 + 4\theta^2\varphi + 6\theta^2\varphi^2 + 4\theta^2\varphi^3 + \theta^2\varphi^4$

RP: $1 + 2x + x^2$

$\chi: 1 - 2x + x^2$

DP: $\varphi^4 + \theta\varphi^2$

Möbius Function:

	{c,d,e,f}	{a,c,d,e,f}	{b,c,d,e,f}	{a,b,c,d,e,f}
{c,d,e,f}	1	-1	-1	1
{a,c,d,e,f}	0	1	0	-1
{b,c,d,e,f}	0	0	1	-1
{a,b,c,d,e,f}	0	0	0	1

Matroid: 6-8

Basis: {a,b}, {a,c}

Rank: 2   Permutations: 60   Augmented: Y   Extended: Y

RG: $2 + 7\varphi + 9\varphi^2 + 5\varphi^3 + \varphi^4 + 3\theta + 10\theta\varphi + 12\theta\varphi^2 + 6\theta\varphi^3 + \theta\varphi^4 + \theta^2 + 3\theta^2\varphi + 3\theta^2\varphi^2 + \theta^2\varphi^3$

RP: $1 + 2x + x^2$

$\chi: 1 - 2x + x^2$

DP: $\varphi^3 + \theta\varphi^2 + \theta^2\varphi$

Möbius Function:

	{d,e,f}	{a,d,e,f}	{b,c,d,e,f}	{a,b,c,d,e,f}
{d,e,f}	1	-1	-1	1
{a,d,e,f}	0	1	0	-1
{b,c,d,e,f}	0	0	1	-1
{a,b,c,d,e,f}	0	0	0	1

Matroid: 6-9

Basis: {a,b}, {a,c}, {a,d}

Rank: 2   Permutations: 60   Augmented: Y   Extended: Y

RG: $3 + 9\varphi + 10\varphi^2 + 5\varphi^3 + \varphi^4 + 4\theta + 11\theta\varphi + 11\theta\varphi^2 + 5\theta\varphi^3 + \theta\varphi^4 + \theta^2 + 2\theta^2\varphi + \theta^2\varphi^2$

RP: $1 + 2x + x^2$

$\chi: 1 - 2x + x^2$

DP: $\varphi^2 + \theta\varphi^3 + \theta^3\varphi$

Möbius Function:

	{e,f}	{a,e,f}	{b,c,d,e,f}	{a,b,c,d,e,f}
{e,f}	1	-1	-1	1
{a,e,f}	0	1	0	-1
{b,c,d,e,f}	0	0	1	-1
{a,b,c,d,e,f}	0	0	0	1

Matroid: 6-10

Basis: {a,b}, {a,c}, {a,d}, {a,e}

Rank: 2   Permutations: 30   Augmented: Y   Extended: Y

RG: $4 + 10\varphi + 10\varphi^2 + 5\varphi^3 + \varphi^4 + 5\theta + 11\theta\varphi + 10\theta\varphi^2 + 5\theta\varphi^3 + \theta\varphi^4 + \theta^2 + \theta^2\varphi$

RP: $1 + 2x + x^2$

$\chi : 1 - 2x + x^2$

DP: $\varphi + \theta\varphi^4 + \theta^4\varphi$

Möbius Function:

	{f}	{a,f}	{b,c,d,e,f}	{a,b,c,d,e,f}
{f}	1	-1	-1	1
{a,f}	0	1	0	-1
{b,c,d,e,f}	0	0	1	-1
{a,b,c,d,e,f}	0	0	0	1

Matroid: 6-11

Basis: {a,b}, {a,c}, {a,d}, {a,e}, {a,f}

Rank: 2   Permutations: 6   Augmented: Y   Extended: N

RG: $5 + 10\varphi + 10\varphi^2 + 5\varphi^3 + \varphi^4 + 6\theta + 10\theta\varphi + 10\theta\varphi^2 + 5\theta\varphi^3 + \theta\varphi^4 + \theta^2$

RP: $1 + 2x + x^2$

$\chi : 1 - 2x + x^2$

DP: $\theta\varphi^5 + \theta^5\varphi$

Möbius Function:

	{}	{a}	{b,c,d,e,f}	{a,b,c,d,e,f}
{}	1	-1	-1	1
{a}	0	1	0	-1
{b,c,d,e,f}	0	0	1	-1
{a,b,c,d,e,f}	0	0	0	1

Matroid: 6-12

Basis: {a,b}, {a,c}, {a,d}, {a,e}, {a,f}, {b,c}, {b,d}, {b,e}, {b,f}

Rank: 2   Permutations: 15   Augmented: N   Extended: N

RG: $9 + 16\varphi + 14\varphi^2 + 6\varphi^3 + \varphi^4 + 6\theta + 6\theta\varphi + 4\theta\varphi^2 + \theta\varphi^3 + \theta^2$

RP: $1 + 3x + x^2$

$\chi : 2 - 3x + x^2$

DP: $\theta^2\varphi^4 + \theta^5\varphi^2$

Möbius Function:

	{}	{a}	{b}	{c,d,e,f}	{a,b,c,d,e,f}
{}	1	-1	-1	-1	2
{a}	0	1	0	0	-1
{b}	0	0	1	0	-1
{c,d,e,f}	0	0	0	1	-1
{a,b,c,d,e,f}	0	0	0	0	1

Matroid: 6-13

Basis: {a,b}, {a,c}, {a,d}, {a,e}, {a,f}, {b,c}, {b,d}, {b,e}, {b,f}, {c,d}, {c,e}, {c,f}

Rank: 2   Permutations: 20   Augmented: N   Extended: N

RG: $12 + 19\varphi + 15\varphi^2 + 6\varphi^3 + \varphi^4 + 6\theta + 3\theta\varphi + \theta\varphi^2 + \theta^2$

RP: $1 + 4x + x^2$

$\chi : 3 - 4x + x^2$

DP: $\theta^3\varphi^3 + \theta^5\varphi^3$

Möbius Function:

	{}	{a}	{b}	{c}	{d,e,f}	{a,b,c,d,e,f}
{}	1	-1	-1	-1	-1	3
{a}	0	1	0	0	0	-1
{b}	0	0	1	0	0	-1
{c}	0	0	0	1	0	-1
{d,e,f}	0	0	0	0	1	-1
{a,b,c,d,e,f}	0	0	0	0	0	1

Matroid: 6-14

Basis: {a,b}, {a,c}, {a,d}, {a,e}, {a,f}, {b,c}, {b,d}, {b,e}, {b,f}, {c,d}, {c,e}, {c,f}, {d,e}, {d,f}

Rank: 2   Permutations: 15   Augmented: N   Extended: N

RG: $14 + 20\varphi + 15\varphi^2 + 6\varphi^3 + \varphi^4 + 6\theta + \theta\varphi + \theta^2$

RP: $1 + 5x + x^2$

$\chi : 4 - 5x + x^2$

DP: $\theta^4\varphi^2 + \theta^5\varphi^4$

Möbius Function:

	{}	{a}	{b}	{c}	{d}	{e,f}	{a,b,c,d,e,f}
{}	1	-1	-1	-1	-1	-1	4
{a}	0	1	0	0	0	0	-1
{b}	0	0	1	0	0	0	-1
{c}	0	0	0	1	0	0	-1
{d}	0	0	0	0	1	0	-1
{e,f}	0	0	0	0	0	1	-1
{a,b,c,d,e,f}	0	0	0	0	0	0	1

Matroid: 6-15

Basis: {a,b}, {a,c}, {a,d}, {a,e}, {a,f}, {b,c}, {b,d}, {b,e}, {b,f}, {c,d}, {c,e}, {c,f}, {d,e}, {d,f}, {e,f}

Rank: 2   Permutations: 1   Augmented: N   Extended: N

RG: $15 + 20\varphi + 15\varphi^2 + 6\varphi^3 + \varphi^4 + 6\theta + \theta^2$

RP: $1 + 6x + x^2$

$\chi : 5 - 6x + x^2$

DP: $\theta^5\varphi^6$

Möbius Function:

	{}	{a}	{b}	{c}	{d}	{e}	{f}	{a,b,c,d,e,f}
{}	1	-1	-1	-1	-1	-1	-1	5
{a}	0	1	0	0	0	0	0	-1
{b}	0	0	1	0	0	0	0	-1
{c}	0	0	0	1	0	0	0	-1
{d}	0	0	0	0	1	0	0	-1
{e}	0	0	0	0	0	1	0	-1
{f}	0	0	0	0	0	0	1	-1
{a,b,c,d,e,f}	0	0	0	0	0	0	0	1

Matroid: 6-16

Basis: {a,b}, {a,c}, {a,d}, {a,e}, {a,f}, {b,c}, {b,d}, {b,e}, {b,f}, {c,d}, {c,e}, {d,f}, {e,f}

Rank: 2   Permutations: 45   Augmented: N   Extended: N

RG: $13 + 20\varphi + 15\varphi^2 + 6\varphi^3 + \varphi^4 + 6\theta + 2\theta\varphi + \theta^2$

RP: $1 + 4x + x^2$

$\chi : 3 - 4x + x^2$

DP: $2\theta^4\varphi^2 + \theta^5\varphi^2$

Möbius Function:

	{}	{a}	{b}	{c,f}	{d,e}	{a,b,c,d,e,f}
{}	1	-1	-1	-1	-1	3
{a}	0	1	0	0	0	-1
{b}	0	0	1	0	0	-1
{c,f}	0	0	0	1	0	-1
{d,e}	0	0	0	0	1	-1
{a,b,c,d,e,f}	0	0	0	0	0	1

Matroid: 6-17

Basis: {a,b}, {a,c}, {a,d}, {a,e}, {a,f}, {b,c}, {b,d}, {b,e}, {c,f}, {d,f}, {e,f}

Rank: 2   Permutations: 60   Augmented: N   Extended: N

RG: $11 + 19\varphi + 15\varphi^2 + 6\varphi^3 + \varphi^4 + 6\theta + 4\theta\varphi + \theta\varphi^2 + \theta^2$

RP: $1 + 3x + x^2$

$\chi : 2 - 3x + x^2$

DP: $\theta^3\varphi^3 + \theta^4\varphi^2 + \theta^5\varphi$

Möbius Function:

	{}	{a}	{b,f}	{c,d,e}	{a,b,c,d,e,f}
{}	1	-1	-1	-1	2
{a}	0	1	0	0	-1
{b,f}	0	0	1	0	-1
{c,d,e}	0	0	0	1	-1
{a,b,c,d,e,f}	0	0	0	0	1

Matroid: 6-18

Basis: {a,b}, {a,c}, {a,d}, {a,e}, {b,c}, {b,d}, {b,e}

Rank: 2   Permutations: 60   Augmented: N   Extended: Y

RG: $7 + 16\varphi + 14\varphi^2 + 6\varphi^3 + \varphi^4 + 5\theta + 8\theta\varphi + 4\theta\varphi^2 + \theta\varphi^3 + \theta^2 + \theta^2\varphi$

RP: $1 + 3x + x^2$

$\chi : 2 - 3x + x^2$

DP: $\varphi + \theta^2\varphi^3 + \theta^4\varphi^2$

Möbius Function:

	{f}	{a,f}	{b,f}	{c,d,e,f}	{a,b,c,d,e,f}
{f}	1	-1	-1	-1	2
{a,f}	0	1	0	0	-1
{b,f}	0	0	1	0	-1
{c,d,e,f}	0	0	0	1	-1
{a,b,c,d,e,f}	0	0	0	0	1

Matroid: 6-19

Basis: {a,b}, {a,c}, {a,d}, {a,e}, {b,c}, {b,d}, {b,e}, {c,d}, {c,e}

Rank: 2   Permutations: 60   Augmented: N   Extended: Y

RG: $9 + 19\varphi + 15\varphi^2 + 6\varphi^3 + \varphi^4 + 5\theta + 6\theta\varphi + \theta\varphi^2 + \theta^2 + \theta^2\varphi$

RP: $1 + 4x + x^2$

$\chi : 3 - 4x + x^2$

DP: $\varphi + \theta^3\varphi^2 + \theta^4\varphi^3$

Möbius Function:

	{f}	{a,f}	{b,f}	{c,f}	{d,e,f}	{a,b,c,d,e,f}
{f}	1	-1	-1	-1	-1	3
{a,f}	0	1	0	0	0	-1
{b,f}	0	0	1	0	0	-1
{c,f}	0	0	0	1	0	-1
{d,e,f}	0	0	0	0	1	-1
{a,b,c,d,e,f}	0	0	0	0	0	1

Matroid: 6-20

Basis: {a,b}, {a,c}, {a,d}, {a,e}, {b,c}, {b,d}, {b,e}, {c,d}, {c,e}, {d,e}

Rank: 2   Permutations: 6   Augmented: N   Extended: Y

RG: $10 + 20\varphi + 15\varphi^2 + 6\varphi^3 + \varphi^4 + 5\theta + 5\theta\varphi + \theta^2 + \theta^2\varphi$

RP: $1 + 5x + x^2$

$\chi : 4 - 5x + x^2$

DP: $\varphi + \theta^4\varphi^5$

Möbius Function:

	{f}	{a,f}	{b,f}	{c,f}	{d,f}	{e,f}	{a,b,c,d,e,f}
{f}	1	-1	-1	-1	-1	-1	4
{a,f}	0	1	0	0	0	0	-1
{b,f}	0	0	1	0	0	0	-1
{c,f}	0	0	0	1	0	0	-1
{d,f}	0	0	0	0	1	0	-1
{e,f}	0	0	0	0	0	1	-1
{a,b,c,d,e,f}	0	0	0	0	0	0	1

Matroid: 6-21

Basis: {a,b}, {a,c}, {a,d}, {a,e}, {b,c}, {b,d}, {b,f}, {c,e}, {c,f}, {d,e}, {d,f}, {e,f}

Rank: 2　　　　Permutations: 15　　　　Augmented: N　　　　Extended: N

RG: $12 + 20\varphi + 15\varphi^2 + 6\varphi^3 + \varphi^4 + 6\theta + 3\theta\varphi + \theta^2$

RP: $1 + 3x + x^2$

$\chi : 2 - 3x + x^2$

DP: $3\theta^4\varphi^2$

Möbius Function:

	{}	{a,f}	{b,e}	{c,d}	{a,b,c,d,e,f}
{}	1	-1	-1	-1	2
{a,f}	0	1	0	0	-1
{b,e}	0	0	1	0	-1
{c,d}	0	0	0	1	-1
{a,b,c,d,e,f}	0	0	0	0	1

Matroid: 6-22

Basis: {a,b}, {a,c}, {a,d}, {a,e}, {b,c}, {b,d}, {c,e}, {d,e}

Rank: 2　　　　Permutations: 90　　　　Augmented: N　　　　Extended: Y

RG: $8 + 18\varphi + 15\varphi^2 + 6\varphi^3 + \varphi^4 + 5\theta + 7\theta\varphi + 2\theta\varphi^2 + \theta^2 + \theta^2\varphi$

RP: $1 + 3x + x^2$

$\chi : 2 - 3x + x^2$

DP: $\varphi + 2\theta^3\varphi^2 + \theta^4\varphi$

Möbius Function:

	{f}	{a,f}	{b,e,f}	{c,d,f}	{a,b,c,d,e,f}
{f}	1	-1	-1	-1	2
{a,f}	0	1	0	0	-1
{b,e,f}	0	0	1	0	-1
{c,d,f}	0	0	0	1	-1
{a,b,c,d,e,f}	0	0	0	0	1

Matroid: 6-23

Basis: {a,b}, {a,c}, {a,d}, {a,e}, {b,f}, {c,f}, {d,f}, {e,f}

Rank: 2　　　　Permutations: 15　　　　Augmented: N　　　　Extended: N

RG: $8 + 16\varphi + 14\varphi^2 + 6\varphi^3 + \varphi^4 + 6\theta + 7\theta\varphi + 4\theta\varphi^2 + \theta\varphi^3 + \theta^2$

RP: $1 + 2x + x^2$

$\chi : 1 - 2x + x^2$

DP: $\theta^2\varphi^4 + \theta^4\varphi^2$

Möbius Function:

	{}	{a,f}	{b,c,d,e}	{a,b,c,d,e,f}
{}	1	-1	-1	1
{a,f}	0	1	0	-1
{b,c,d,e}	0	0	1	-1
{a,b,c,d,e,f}	0	0	0	1

Matroid: 6-24

Basis: {a,b}, {a,c}, {a,d}, {b,c}, {b,d}

Rank: 2   Permutations: 90   Augmented: N   Extended: Y

RG: $5 + 14\varphi + 14\varphi^2 + 6\varphi^3 + \varphi^4 + 4\theta + 9\theta\varphi + 6\theta\varphi^2 + \theta\varphi^3 + \theta^2 + 2\theta^2\varphi + \theta^2\varphi^2$

RP: $1 + 3x + x^2$

$\chi : 2 - 3x + x^2$

DP: $\varphi^2 + \theta^2\varphi^2 + \theta^3\varphi^2$

Möbius Function:

	{e,f}	{a,e,f}	{b,e,f}	{c,d,e,f}	{a,b,c,d,e,f}
{e,f}	1	-1	-1	-1	2
{a,e,f}	0	1	0	0	-1
{b,e,f}	0	0	1	0	-1
{c,d,e,f}	0	0	0	1	-1
{a,b,c,d,e,f}	0	0	0	0	1

Matroid: 6-25

Basis: {a,b}, {a,c}, {a,d}, {b,c}, {b,d}, {c,d}

Rank: 2   Permutations: 15   Augmented: N   Extended: Y

RG: $6 + 16\varphi + 15\varphi^2 + 6\varphi^3 + \varphi^4 + 4\theta + 8\theta\varphi + 4\theta\varphi^2 + \theta^2 + 2\theta^2\varphi + \theta^2\varphi^2$

RP: $1 + 4x + x^2$

$\chi : 3 - 4x + x^2$

DP: $\varphi^2 + \theta^3\varphi^4$

Möbius Function:

	{e,f}	{a,e,f}	{b,e,f}	{c,e,f}	{d,e,f}	{a,b,c,d,e,f}
{e,f}	1	-1	-1	-1	-1	3
{a,e,f}	0	1	0	0	0	-1
{b,e,f}	0	0	1	0	0	-1
{c,e,f}	0	0	0	1	0	-1
{d,e,f}	0	0	0	0	1	-1
{a,b,c,d,e,f}	0	0	0	0	0	1

Matroid: 6-26

Basis: {a,b}, {a,c}, {a,d}, {b,e}, {b,f}, {c,e}, {c,f}, {d,e}, {d,f}

Rank: 2   Permutations: 10   Augmented: N   Extended: N

RG: $9 + 18\varphi + 15\varphi^2 + 6\varphi^3 + \varphi^4 + 6\theta + 6\theta\varphi + 2\theta\varphi^2 + \theta^2$

RP: $1 + 2x + x^2$

$\chi : 1 - 2x + x^2$

DP: $2\theta^3\varphi^3$

Möbius Function:

	{}	{a,e,f}	{b,c,d}	{a,b,c,d,e,f}
{}	1	-1	-1	1
{a,e,f}	0	1	0	-1
{b,c,d}	0	0	1	-1
{a,b,c,d,e,f}	0	0	0	1

Matroid: 6-27

Basis: {a,b}, {a,c}, {a,d}, {b,e}, {c,e}, {d,e}

Rank: 2   Permutations: 60   Augmented: N   Extended: Y

RG: $6 + 15\varphi + 14\varphi^2 + 6\varphi^3 + \varphi^4 + 5\theta + 9\theta\varphi + 5\theta\varphi^2 + \theta\varphi^3 + \theta^2 + \theta^2\varphi$

RP: $1 + 2x + x^2$

$\chi : 1 - 2x + x^2$

DP: $\varphi + \theta^2\varphi^3 + \theta^3\varphi^2$

Möbius Function:

	{f}	{a,e,f}	{b,c,d,f}	{a,b,c,d,e,f}
{f}	1	-1	-1	1
{a,e,f}	0	1	0	-1
{b,c,d,f}	0	0	1	-1
{a,b,c,d,e,f}	0	0	0	1

Matroid: 6-28

Basis: {a,b}, {a,c}, {b,c}

Rank: 2   Permutations: 20   Augmented: N   Extended: Y

RG: $3 + 10\varphi + 12\varphi^2 + 6\varphi^3 + \varphi^4 + 3\theta + 9\theta\varphi + 9\theta\varphi^2 + 3\theta\varphi^3 + \theta^2 + 3\theta^2\varphi + 3\theta^2\varphi^2 + \theta^2\varphi^3$

RP: $1 + 3x + x^2$

$\chi : 2 - 3x + x^2$

DP: $\varphi^3 + \theta^2\varphi^3$

Möbius Function:

	{d,e,f}	{a,d,e,f}	{b,d,e,f}	{c,d,e,f}	{a,b,c,d,e,f}
{d,e,f}	1	-1	-1	-1	2
{a,d,e,f}	0	1	0	0	-1
{b,d,e,f}	0	0	1	0	-1
{c,d,e,f}	0	0	0	1	-1
{a,b,c,d,e,f}	0	0	0	0	1

Matroid: 6-29

Basis: {a,b}, {a,c}, {b,d}, {c,d}

Rank: 2   Permutations: 45   Augmented: N   Extended: Y

RG: $4 + 12\varphi + 13\varphi^2 + 6\varphi^3 + \varphi^4 + 4\theta + 10\theta\varphi + 8\theta\varphi^2 + 2\theta\varphi^3 + \theta^2 + 2\theta^2\varphi + \theta^2\varphi^2$

RP: $1 + 2x + x^2$

$\chi: 1 - 2x + x^2$

DP: $\varphi^2 + 2\theta^2\varphi^2$

Möbius Function:

	{e,f}	{a,d,e,f}	{b,c,e,f}	{a,b,c,d,e,f}
{e,f}	1	-1	-1	1
{a,d,e,f}	0	1	0	-1
{b,c,e,f}	0	0	1	-1
{a,b,c,d,e,f}	0	0	0	1

Matroid: 6-30

Basis: {a,b,c}

Rank: 3   Permutations: 20   Augmented: Y   Extended: Y

RG: $1 + 3\varphi + 3\varphi^2 + \varphi^3 + 3\theta + 9\theta\varphi + 9\theta\varphi^2 + 3\theta\varphi^3 + 3\theta^2 + 9\theta^2\varphi + 9\theta^2\varphi^2 + 3\theta^2\varphi^3 + \theta^3 + 3\theta^3\varphi + 3\theta^3\varphi^2 + \theta^3\varphi^3$

RP: $1 + 3x + 3x^2 + x^3$

$\chi: -1 + 3x - 3x^2 + x^3$

DP: $\varphi^3 + \theta\varphi^3$

Möbius Function:

	{d,e,f}	{a,d,e,f}	{b,d,e,f}	{c,d,e,f}	{a,b,d,e,f}	{a,c,d,e,f}	{b,c,d,e,f}	{a,b,c,d,e,f}
{d,e,f}	1	-1	-1	-1	1	1	1	-1
{a,d,e,f}	0	1	0	0	-1	-1	0	1
{b,d,e,f}	0	0	1	0	-1	0	-1	1
{c,d,e,f}	0	0	0	1	0	-1	-1	1
{a,b,d,e,f}	0	0	0	0	1	0	0	-1
{a,c,d,e,f}	0	0	0	0	0	1	0	-1
{b,c,d,e,f}	0	0	0	0	0	0	1	-1
{a,b,c,d,e,f}	0	0	0	0	0	0	0	1

Matroid: 6-31
Basis: {a,b,c}, {a,b,d}
Rank: 3   Permutations: 90   Augmented: Y   Extended: Y
RG: $2 + 5\varphi + 4\varphi^2 + \varphi^3 + 5\theta + 12\theta\varphi + 9\theta\varphi^2 + 2\theta\varphi^3 + 4\theta^2 + 9\theta^2\varphi + 6\theta^2\varphi^2 + \theta^2\varphi^3 + \theta^3 + 2\theta^3\varphi + \theta^3\varphi^2$
RP: $1 + 3x + 3x^2 + x^3$
$\chi : -1 + 3x - 3x^2 + x^3$
DP: $\varphi^2 + \theta\varphi^2 + \theta^2\varphi^2$
Möbius Function:

	{e,f}	{a,e,f}	{b,e,f}	{a,b,e,f}	{c,d,e,f}	{a,c,d,e,f}	{b,c,d,e,f}	{a,b,c,d,e,f}
{e,f}	1	-1	-1	1	-1	1	1	-1
{a,e,f}	0	1	0	-1	0	-1	0	1
{b,e,f}	0	0	1	-1	0	0	-1	1
{a,b,e,f}	0	0	0	1	0	0	0	-1
{c,d,e,f}	0	0	0	0	1	-1	-1	1
{a,c,d,e,f}	0	0	0	0	0	1	0	-1
{b,c,d,e,f}	0	0	0	0	0	0	1	-1
{a,b,c,d,e,f}	0	0	0	0	0	0	0	1

Matroid: 6-32
Basis: {a,b,c}, {a,b,d}, {a,b,e}
Rank: 3   Permutations: 60   Augmented: Y   Extended: Y
RG: $3 + 6\varphi + 4\varphi^2 + \varphi^3 + 7\theta + 13\theta\varphi + 8\theta\varphi^2 + 2\theta\varphi^3 + 5\theta^2 + 8\theta^2\varphi + 4\theta^2\varphi^2 + \theta^2\varphi^3 + \theta^3 + \theta^3\varphi$
RP: $1 + 3x + 3x^2 + x^3$
$\chi : -1 + 3x - 3x^2 + x^3$
DP: $\varphi + \theta\varphi^3 + \theta^3\varphi^2$

Matroid: 6-33
Basis: {a,b,c}, {a,b,d}, {a,b,e}, {a,b,f}
Rank: 3   Permutations: 15   Augmented: Y   Extended: N
RG: $4 + 6\varphi + 4\varphi^2 + \varphi^3 + 9\theta + 12\theta\varphi + 8\theta\varphi^2 + 2\theta\varphi^3 + 6\theta^2 + 6\theta^2\varphi + 4\theta^2\varphi^2 + \theta^2\varphi^3 + \theta^3$
RP: $1 + 3x + 3x^2 + x^3$
$\chi : -1 + 3x - 3x^2 + x^3$
DP: $\theta\varphi^4 + \theta^4\varphi^2$

Möbius Function:

	{}	{a}	{b}	{a,b}	{c,d,e,f}	{a,c,d,e,f}	{b,c,d,e,f}	{a,b,c,d,e,f}
{}	1	-1	-1	1	-1	1	1	-1
{a}	0	1	0	-1	0	-1	0	1
{b}	0	0	1	-1	0	0	-1	1
{a,b}	0	0	0	1	0	0	0	-1
{c,d,e,f}	0	0	0	0	1	-1	-1	1
{a,c,d,e,f}	0	0	0	0	0	1	0	-1
{b,c,d,e,f}	0	0	0	0	0	0	1	-1
{a,b,c,d,e,f}	0	0	0	0	0	0	0	1

Matroid: 6-34

Basis: {a,b,c}, {a,b,d}, {a,b,e}, {a,b,f}, {a,c,d}, {a,c,e}, {a,c,f}

Rank: 3   Permutations: 60   Augmented: Y   Extended: N

RG: $7 + 9\varphi + 5\varphi^2 + \varphi^3 + 12\theta + 12\theta\varphi + 6\theta\varphi^2 + \theta\varphi^3 + 6\theta^2 + 3\theta^2\varphi + \theta^2\varphi^2 + \theta^3$

RP: $1 + 4x + 4x^2 + x^3$

$\chi : -2 + 5x - 4x^2 + x^3$

DP: $\theta^2\varphi^3 + \theta^4\varphi^2 + \theta^7\varphi$

Möbius Function:

	{}	{a}	{b}	{c}	{a,b}	{a,c}	{d,e,f}	{a,d,e,f}	{b,c,d,e,f}	{a,b,c,d,e,f}
{}	1	-1	-1	-1	1	1	-1	1	2	-2
{a}	0	1	0	0	-1	-1	0	-1	0	2
{b}	0	0	1	0	-1	0	0	0	-1	1
{c}	0	0	0	1	0	-1	0	0	-1	1
{a,b}	0	0	0	0	1	0	0	0	0	-1
{a,c}	0	0	0	0	0	1	0	0	0	-1
{d,e,f}	0	0	0	0	0	0	1	-1	-1	1
{a,d,e,f}	0	0	0	0	0	0	0	1	0	-1
{b,c,d,e,f}	0	0	0	0	0	0	0	0	1	-1
{a,b,c,d,e,f}	0	0	0	0	0	0	0	0	0	1

Matroid: 6-35

Basis: {a,b,c}, {a,b,d}, {a,b,e}, {a,b,f}, {a,c,d}, {a,c,e}, {a,c,f}, {a,d,e}, {a,d,f}

Rank: 3   Permutations: 60   Augmented: Y   Extended: N

RG: $9 + 10\varphi + 5\varphi^2 + \varphi^3 + 14\theta + 11\theta\varphi + 5\theta\varphi^2 + \theta\varphi^3 + 6\theta^2 + \theta^2\varphi + \theta^3$

RP: $1 + 5x + 5x^2 + x^3$

$\chi : -3 + 7x - 5x^2 + x^3$

DP: $\theta^3\varphi^2 + \theta^4\varphi^3 + \theta^9\varphi$

Möbius Function:

	{}	{a}	{b}	{c}	{d}	{a,b}	{a,c}	{a,d}	{e,f}	{a,e,f}	{b,c,d,e,f}	{a,b,c,d,e,f}
{}	1	-1	-1	-1	-1	1	1	1	-1	1	3	-3
{a}	0	1	0	0	0	-1	-1	-1	0	-1	0	3
{b}	0	0	1	0	0	-1	0	0	0	0	-1	1
{c}	0	0	0	1	0	0	-1	0	0	0	-1	1
{d}	0	0	0	0	1	0	0	-1	0	0	-1	1
{a,b}	0	0	0	0	0	1	0	0	0	0	0	-1
{a,c}	0	0	0	0	0	0	1	0	0	0	0	-1
{a,d}	0	0	0	0	0	0	0	1	0	0	0	-1
{e,f}	0	0	0	0	0	0	0	0	1	-1	-1	1
{a,e,f}	0	0	0	0	0	0	0	0	0	1	0	-1
{b,c,d,e,f}	0	0	0	0	0	0	0	0	0	0	1	-1
{a,b,c,d,e,f}	0	0	0	0	0	0	0	0	0	0	0	1

Matroid: 6-36

Basis: {a,b,c}, {a,b,d}, {a,b,e}, {a,b,f}, {a,c,d}, {a,c,e}, {a,c,f}, {a,d,e}, {a,d,f}, {a,e,f}

Rank: 3   Permutations: 6   Augmented: Y   Extended: N

RG: $10 + 10\varphi + 5\varphi^2 + \varphi^3 + 15\theta + 10\theta\varphi + 5\theta\varphi^2 + \theta\varphi^3 + 6\theta^2 + \theta^3$

RP: $1 + 6x + 6x^2 + x^3$

χ : $-4 + 9x - 6x^2 + x^3$

DP: $\theta^4\varphi^5 + \theta^{10}\varphi$

Matroid: 6-37

Basis: {a,b,c}, {a,b,d}, {a,b,e}, {a,b,f}, {a,c,d}, {a,c,e}, {a,c,f}, {a,d,e}, {a,d,f}, {a,e,f}, {b,c,d}, {b,c,e}, {b,c,f}, {b,d,e}, {b,d,f}, {b,e,f}

Rank: 3   Permutations: 15   Augmented: N   Extended: N

RG: $16 + 14\varphi + 6\varphi^2 + \varphi^3 + 15\theta + 4\theta\varphi + \theta\varphi^2 + 6\theta^2 + \theta^3$

RP: $1 + 10x + 6x^2 + x^3$

χ : $-7 + 12x - 6x^2 + x^3$

DP: $\theta^7\varphi^4 + \theta^{10}\varphi^2$

Matroid: 6-38

Basis: {a,b,c}, {a,b,d}, {a,b,e}, {a,b,f}, {a,c,d}, {a,c,e}, {a,c,f}, {a,d,e}, {a,d,f}, {a,e,f}, {b,c,d}, {b,c,e}, {b,c,f}, {b,d,e}, {b,d,f}, {b,e,f}, {c,d,e}, {c,d,f}, {c,e,f}

Rank: 3   Permutations: 20   Augmented: N   Extended: N

RG: $19 + 15\varphi + 6\varphi^2 + \varphi^3 + 15\theta + \theta\varphi + 6\theta^2 + \theta^3$

RP: $1 + 13x + 6x^2 + x^3$

χ : $-9 + 14x - 6x^2 + x^3$

DP: $\theta^9\varphi^3 + \theta^{10}\varphi^3$

Matroid: 6-39

Basis: {a,b,c}, {a,b,d}, {a,b,e}, {a,b,f}, {a,c,d}, {a,c,e}, {a,c,f}, {a,d,e}, {a,d,f}, {a,e,f}, {b,c,d}, {b,c,e}, {b,c,f}, {b,d,e}, {b,d,f}, {b,e,f}, {c,d,e}, {c,d,f}, {c,e,f}, {d,e,f}

Rank: 3   Permutations: 1   Augmented: N   Extended: N

RG: $20 + 15\varphi + 6\varphi^2 + \varphi^3 + 15\theta + 6\theta^2 + \theta^3$

RP: $1 + 15x + 6x^2 + x^3$

$\chi : -10 + 15x - 6x^2 + x^3$

DP: $\theta^{10}\varphi^6$

Matroid: 6-40

Basis: {a,b,c}, {a,b,d}, {a,b,e}, {a,b,f}, {a,c,d}, {a,c,e}, {a,c,f}, {a,d,e}, {a,d,f}, {a,e,f}, {b,c,d}, {b,c,e}, {b,c,f}, {b,d,e}, {b,d,f}, {c,d,e}, {c,e,f}, {d,e,f}

Rank: 3   Permutations: 90   Augmented: N   Extended: N

RG: $18 + 15\varphi + 6\varphi^2 + \varphi^3 + 15\theta + 2\theta\varphi + 6\theta^2 + \theta^3$

RP: $1 + 11x + 6x^2 + x^3$

$\chi : -8 + 13x - 6x^2 + x^3$

DP: $\theta^8\varphi + 2\theta^9\varphi^2 + \theta^{10}\varphi$

Matroid: 6-41

Basis: {a,b,c}, {a,b,d}, {a,b,e}, {a,b,f}, {a,c,d}, {a,c,e}, {a,c,f}, {a,d,e}, {a,d,f}, {b,c,d}, {b,c,e}, {b,c,f}, {b,d,e}, {b,d,f}

Rank: 3   Permutations: 90   Augmented: N   Extended: N

RG: $14 + 14\varphi + 6\varphi^2 + \varphi^3 + 14\theta + 6\theta\varphi + \theta\varphi^2 + 6\theta^2 + \theta^2\varphi + \theta^3$

RP: $1 + 8x + 5x^2 + x^3$

$\chi : -5 + 9x - 5x^2 + x^3$

DP: $\theta^5\varphi^2 + \theta^7\varphi^2 + \theta^9\varphi^2$

Matroid: 6-42

Basis: {a,b,c}, {a,b,d}, {a,b,e}, {a,b,f}, {a,c,d}, {a,c,e}, {a,c,f}, {a,d,e}, {a,d,f}, {b,c,d}, {b,c,e}, {b,c,f}, {b,d,e}, {b,d,f}, {c,d,e}, {c,d,f}

Rank: 3   Permutations: 15   Augmented: N   Extended: N

RG: $16 + 15\varphi + 6\varphi^2 + \varphi^3 + 14\theta + 4\theta\varphi + 6\theta^2 + \theta^2\varphi + \theta^3$

RP: $1 + 10x + 5x^2 + x^3$

$\chi : -6 + 10x - 5x^2 + x^3$

DP: $\theta^6\varphi^2 + \theta^9\varphi^4$

Matroid: 6-43

Basis: {a,b,c}, {a,b,d}, {a,b,e}, {a,b,f}, {a,c,d}, {a,c,e}, {a,c,f}, {a,d,e}, {a,d,f}, {b,c,d}, {b,c,e}, {b,c,f}, {b,d,e}, {b,e,f}, {c,d,f}, {c,e,f}, {d,e,f}

Rank: 3   Permutations: 120   Augmented: N   Extended: N

RG: $17 + 15\varphi + 6\varphi^2 + \varphi^3 + 15\theta + 3\theta\varphi + 6\theta^2 + \theta^3$

RP: $1 + 9x + 6x^2 + x^3$

χ : $-7 + 12x - 6x^2 + x^3$

DP: $3\theta^8\varphi + 3\theta^9\varphi$

Matroid: 6-44

Basis: {a,b,c}, {a,b,d}, {a,b,e}, {a,b,f}, {a,c,d}, {a,c,e}, {a,c,f}, {a,d,e}, {a,d,f}, {b,c,e}, {b,c,f}, {b,d,e}, {b,d,f}, {b,e,f}, {c,d,e}, {c,d,f}, {c,e,f}, {d,e,f}

Rank: 3   Permutations: 10   Augmented: N   Extended: N

RG: $18 + 15\varphi + 6\varphi^2 + \varphi^3 + 15\theta + 2\theta\varphi + 6\theta^2 + \theta^3$

RP: $1 + 11x + 6x^2 + x^3$

χ : $-8 + 13x - 6x^2 + x^3$

DP: $2\theta^9\varphi^3$

Matroid: 6-45

Basis: {a,b,c}, {a,b,d}, {a,b,e}, {a,b,f}, {a,c,d}, {a,c,e}, {a,c,f}, {a,d,e}, {a,d,f}, {b,c,e}, {b,c,f}, {b,d,e}, {b,d,f}, {c,d,e}, {c,d,f}

Rank: 3   Permutations: 60   Augmented: N   Extended: N

RG: $15 + 15\varphi + 6\varphi^2 + \varphi^3 + 14\theta + 5\theta\varphi + 6\theta^2 + \theta^2\varphi + \theta^3$

RP: $1 + 8x + 5x^2 + x^3$

χ: $-5 + 9x - 5x^2 + x^3$

DP: $\theta^6\varphi^2 + \theta^8\varphi^3 + \theta^9\varphi$

Matroid: 6-46

Basis: {a,b,c}, {a,b,d}, {a,b,e}, {a,b,f}, {a,c,d}, {a,c,e}, {a,c,f}, {a,d,e}, {a,d,f}, {b,c,e}, {b,d,e}, {b,e,f}, {c,d,e}, {c,e,f}, {d,e,f}

Rank: 3   Permutations: 60   Augmented: N   Extended: N

RG: $15 + 14\varphi + 6\varphi^2 + \varphi^3 + 15\theta + 5\theta\varphi + \theta\varphi^2 + 6\theta^2 + \theta^3$

RP: $1 + 8x + 6x^2 + x^3$

χ: $-6 + 11x - 6x^2 + x^3$

DP: $\theta^6\varphi + \theta^7\varphi^3 + \theta^9\varphi^2$

Matroid: 6-47

Basis: {a,b,c}, {a,b,d}, {a,b,e}, {a,b,f}, {a,c,d}, {a,c,e}, {a,c,f}, {b,c,d}, {b,c,e}, {b,c,f}

Rank: 3   Permutations: 20   Augmented: N   Extended: N

RG: $10 + 12\varphi + 6\varphi^2 + \varphi^3 + 12\theta + 9\theta\varphi + 3\theta\varphi^2 + 6\theta^2 + 3\theta^2\varphi + \theta^2\varphi^2 + \theta^3$

RP: $1 + 6x + 4x^2 + x^3$

$\chi: -3 + 6x - 4x^2 + x^3$

DP: $\theta^3\varphi^3 + \theta^7\varphi^3$

Matroid: 6-48

Basis: {a,b,c}, {a,b,d}, {a,b,e}, {a,b,f}, {a,c,d}, {a,c,e}, {a,c,f}, {b,c,d}, {b,c,e}, {b,d,f}, {b,e,f}, {c,d,f}, {c,e,f}

Rank: 3   Permutations: 180   Augmented: N   Extended: N

RG: $13 + 14\varphi + 6\varphi^2 + \varphi^3 + 14\theta + 7\theta\varphi + \theta\varphi^2 + 6\theta^2 + \theta^2\varphi + \theta^3$

RP: $1 + 6x + 5x^2 + x^3$

$\chi: -4 + 8x - 5x^2 + x^3$

DP: $\theta^5\varphi^2 + \theta^6\varphi + \theta^7\varphi + \theta^8\varphi^2$

Matroid: 6-49

Basis: {a,b,c}, {a,b,d}, {a,b,e}, {a,b,f}, {a,c,d}, {a,c,e}, {a,c,f}, {b,c,d}, {b,d,e}, {b,d,f}, {c,d,e}, {c,d,f}

Rank: 3   Permutations: 45   Augmented: N   Extended: N

RG: $12 + 13\varphi + 6\varphi^2 + \varphi^3 + 14\theta + 8\theta\varphi + 2\theta\varphi^2 + 6\theta^2 + \theta^2\varphi + \theta^3$

RP: $1 + 6x + 5x^2 + x^3$

$\chi: -4 + 8x - 5x^2 + x^3$

DP: $\theta^4\varphi^2 + 2\theta^7\varphi^2$

Matroid: 6-50

Basis: {a,b,c}, {a,b,d}, {a,b,e}, {a,b,f}, {a,c,d}, {a,c,e}, {a,d,f}, {a,e,f}

Rank: 3   Permutations: 90   Augmented: Y   Extended: N

RG: $8 + 10\varphi + 5\varphi^2 + \varphi^3 + 13\theta + 12\theta\varphi + 5\theta\varphi^2 + \theta\varphi^3 + 6\theta^2 + 2\theta^2\varphi + \theta^3$

RP: $1 + 4x + 4x^2 + x^3$

$\chi: -2 + 5x - 4x^2 + x^3$

DP: $2\theta^3\varphi^2 + \theta^4\varphi + \theta^8\varphi$

Möbius Function:

	{}	{a}	{b}	{a,b}	{c,f}	{d,e}	{a,c,f}	{a,d,e}	{b,c,d,e,f}	{a,b,c,d,e,f}
{}	1	-1	-1	1	-1	-1	1	1	2	-2
{a}	0	1	0	-1	0	0	-1	-1	0	2
{b}	0	0	1	-1	0	0	0	0	-1	1
{a,b}	0	0	0	1	0	0	0	0	0	-1
{c,f}	0	0	0	0	1	0	-1	0	-1	1
{d,e}	0	0	0	0	0	1	0	-1	-1	1
{a,c,f}	0	0	0	0	0	0	1	0	0	-1
{a,d,e}	0	0	0	0	0	0	0	1	0	-1
{b,c,d,e,f}	0	0	0	0	0	0	0	0	1	-1
{a,b,c,d,e,f}	0	0	0	0	0	0	0	0	0	1

Matroid: 6-51

Basis: {a,b,c}, {a,b,d}, {a,b,e}, {a,b,f}, {a,c,d}, {a,c,e}, {a,d,f}, {a,e,f}, {b,c,d}, {b,c,e}, {b,d,f}, {b,e,f}

Rank: 3   Permutations: 45   Augmented: N   Extended: N

RG: $12 + 14\varphi + 6\varphi^2 + \varphi^3 + 13\theta + 8\theta\varphi + \theta\varphi^2 + 6\theta^2 + 2\theta^2\varphi + \theta^3$

RP: $1 + 6x + 4x^2 + x^3$

$\chi: -3 + 6x - 4x^2 + x^3$

DP: $2\theta^5\varphi^2 + \theta^8\varphi^2$

Matroid: 6-52

Basis: {a,b,c}, {a,b,d}, {a,b,e}, {a,b,f}, {a,c,d}, {a,c,e}, {a,d,f}, {a,e,f}, {b,c,d}, {b,c,f}, {b,d,e}, {b,e,f}, {c,d,e}, {c,d,f}, {c,e,f}, {d,e,f}

Rank: 3   Permutations: 30   Augmented: N   Extended: N

RG: $16 + 15\varphi + 6\varphi^2 + \varphi^3 + 15\theta + 4\theta\varphi + 6\theta^2 + \theta^3$

RP: $1 + 7x + 6x^2 + x^3$

$\chi: -6 + 11x - 6x^2 + x^3$

DP: $6\theta^8\varphi$

Matroid: 6-53

Basis: {a,b,c}, {a,b,d}, {a,b,e}, {a,c,d}, {a,c,e}

Rank: 3   Permutations: 180   Augmented: Y   Extended: Y

RG: $5 + 9\varphi + 5\varphi^2 + \varphi^3 + 9\theta + 14\theta\varphi + 6\theta\varphi^2 + \theta\varphi^3 + 5\theta^2 + 6\theta^2\varphi + \theta^2\varphi^2 + \theta^3 + \theta^3\varphi$

RP: $1 + 4x + 4x^2 + x^3$

$\chi: -2 + 5x - 4x^2 + x^3$

DP: $\varphi + \theta^2\varphi^2 + \theta^3\varphi^2 + \theta^5\varphi$

Matroid: 6-54

Basis: {a,b,c}, {a,b,d}, {a,b,e}, {a,c,d}, {a,c,e}, {a,d,e}

Rank: 3   Permutations: 30   Augmented: Y   Extended: Y

RG: $6 + 10\varphi + 5\varphi^2 + \varphi^3 + 10\theta + 14\theta\varphi + 5\theta\varphi^2 + \theta\varphi^3 + 5\theta^2 + 5\theta^2\varphi + \theta^3 + \theta^3\varphi$

RP: $1 + 5x + 5x^2 + x^3$

χ: $-3 + 7x - 5x^2 + x^3$

DP: $\varphi + \theta^3\varphi^4 + \theta^6\varphi$

Matroid: 6-55

Basis: {a,b,c}, {a,b,d}, {a,b,e}, {a,c,d}, {a,c,e}, {a,d,e}, {b,c,d}, {b,c,e}, {b,d,e}

Rank: 3   Permutations: 60   Augmented: N   Extended: Y

RG: $9 + 14\varphi + 6\varphi^2 + \varphi^3 + 10\theta + 11\theta\varphi + \theta\varphi^2 + 5\theta^2 + 5\theta^2\varphi + \theta^3 + \theta^3\varphi$

RP: $1 + 8x + 5x^2 + x^3$

χ: $-5 + 9x - 5x^2 + x^3$

DP: $\varphi + \theta^5\varphi^3 + \theta^6\varphi^2$

Matroid: 6-56

Basis: {a,b,c}, {a,b,d}, {a,b,e}, {a,c,d}, {a,c,e}, {a,d,e}, {b,c,d}, {b,c,e}, {b,d,e}, {c,d,e}

Rank: 3   Permutations: 6   Augmented: N   Extended: Y

RG: $10 + 15\varphi + 6\varphi^2 + \varphi^3 + 10\theta + 10\theta\varphi + 5\theta^2 + 5\theta^2\varphi + \theta^3 + \theta^3\varphi$

RP: $1 + 10x + 5x^2 + x^3$

χ: $-6 + 10x - 5x^2 + x^3$

DP: $\varphi + \theta^6\varphi^5$

Matroid: 6-57

Basis: {a,b,c}, {a,b,d}, {a,b,e}, {a,c,d}, {a,c,e}, {a,d,e}, {b,c,f}, {b,d,f}, {b,e,f}, {c,d,f}, {c,e,f}, {d,e,f}

Rank: 3   Permutations: 15   Augmented: N   Extended: N

RG: $12 + 14\varphi + 6\varphi^2 + \varphi^3 + 14\theta + 8\theta\varphi + \theta\varphi^2 + 6\theta^2 + \theta^2\varphi + \theta^3$

RP: $1 + 5x + 5x^2 + x^3$

χ: $-3 + 7x - 5x^2 + x^3$

DP: $\theta^6\varphi^2 + \theta^6\varphi^4$

Matroid: 6-58

Basis: {a,b,c}, {a,b,d}, {a,b,e}, {a,c,d}, {a,c,e}, {b,c,d}, {b,c,e}

Rank: 3   Permutations: 60   Augmented: N   Extended: Y

RG: $7 + 12\varphi + 6\varphi^2 + \varphi^3 + 9\theta + 12\theta\varphi + 3\theta\varphi^2 + 5\theta^2 + 6\theta^2\varphi + \theta^2\varphi^2 + \theta^3 + \theta^3\varphi$

RP: $1 + 6x + 4x^2 + x^3$

$\chi: -3 + 6x - 4x^2 + x^3$

DP: $\varphi + \theta^3\varphi^2 + \theta^5\varphi^3$

Matroid: 6-59

Basis: {a,b,c}, {a,b,d}, {a,b,e}, {a,c,d}, {a,c,e}, {b,c,d}, {b,d,e}, {c,d,e}

Rank: 3   Permutations: 90   Augmented: N   Extended: Y

RG: $8 + 13\varphi + 6\varphi^2 + \varphi^3 + 10\theta + 12\theta\varphi + 2\theta\varphi^2 + 5\theta^2 + 5\theta^2\varphi + \theta^3 + \theta^3\varphi$

RP: $1 + 6x + 5x^2 + x^3$

$\chi: -4 + 8x - 5x^2 + x^3$

DP: $\varphi + \theta^4\varphi + 2\theta^5\varphi^2$

Matroid: 6-60

Basis: {a,b,c}, {a,b,d}, {a,b,e}, {a,c,d}, {a,c,e}, {b,c,f}, {b,d,f}, {b,e,f}, {c,d,f}, {c,e,f}

Rank: 3   Permutations: 90   Augmented: N   Extended: N

RG: $10 + 13\varphi + 6\varphi^2 + \varphi^3 + 13\theta + 10\theta\varphi + 2\theta\varphi^2 + 6\theta^2 + 2\theta^2\varphi + \theta^3$

RP: $1 + 4x + 4x^2 + x^3$

$\chi: -2 + 5x - 4x^2 + x^3$

DP: $\theta^4\varphi^2 + \theta^5\varphi^2 + \theta^6\varphi^2$

Möbius Function:

	{}	{b}	{c}	{a,f}	{d,e}	{a,b,f}	{a,c,f}	{a,d,e,f}	{b,c,d,e}	{a,b,c,d,e,f}
{}	1	-1	-1	-1	-1	1	1	1	2	-2
{b}	0	1	0	0	0	-1	0	0	-1	1
{c}	0	0	1	0	0	0	-1	0	-1	1
{a,f}	0	0	0	1	0	-1	-1	-1	0	2
{d,e}	0	0	0	0	1	0	0	-1	-1	1
{a,b,f}	0	0	0	0	0	1	0	0	0	-1
{a,c,f}	0	0	0	0	0	0	1	0	0	-1
{a,d,e,f}	0	0	0	0	0	0	0	1	0	-1
{b,c,d,e}	0	0	0	0	0	0	0	0	1	-1
{a,b,c,d,e,f}	0	0	0	0	0	0	0	0	0	1

Matroid: 6-61

Basis: {a,b,c}, {a,b,d}, {a,b,e}, {a,c,f}, {a,d,f}, {a,e,f}

Rank: 3   Permutations: 60   Augmented: Y   Extended: N

RG: $6 + 9\varphi + 5\varphi^2 + \varphi^3 + 11\theta + 13\theta\varphi + 6\theta\varphi^2 + \theta\varphi^3 + 6\theta^2 + 4\theta^2\varphi + \theta^2\varphi^2 + \theta^3$

RP: $1 + 3x + 3x^2 + x^3$

$\chi: -1 + 3x - 3x^2 + x^3$

DP: $\theta^2\varphi^3 + \theta^3\varphi^2 + \theta^6\varphi$

Möbius Function:

	{}	{a}	{b,f}	{a,b,f}	{c,d,e}	{a,c,d,e}	{b,c,d,e,f}	{a,b,c,d,e,f}
{}	1	-1	-1	1	-1	1	1	-1
{a}	0	1	0	-1	0	-1	0	1
{b,f}	0	0	1	-1	0	0	-1	1
{a,b,f}	0	0	0	1	0	0	0	-1
{c,d,e}	0	0	0	0	1	-1	-1	1
{a,c,d,e}	0	0	0	0	0	1	0	-1
{b,c,d,e,f}	0	0	0	0	0	0	1	-1
{a,b,c,d,e,f}	0	0	0	0	0	0	0	1

Matroid: 6-62

Basis: {a,b,c}, {a,b,d}, {a,b,e}, {a,c,f}, {a,d,f}, {a,e,f}, {b,c,f}, {b,d,f}, {b,e,f}

Rank: 3   Permutations: 20   Augmented: N   Extended: N

RG: $9 + 12\varphi + 6\varphi^2 + \varphi^3 + 12\theta + 10\theta\varphi + 3\theta\varphi^2 + 6\theta^2 + 3\theta^2\varphi + \theta^2\varphi^2 + \theta^3$

RP: $1 + 4x + 4x^2 + x^3$

$\chi: -2 + 5x - 4x^2 + x^3$

DP: $\theta^3\varphi^3 + \theta^6\varphi^3$

Möbius Function:

	{}	{a}	{b}	{f}	{a,b,f}	{c,d,e}	{a,c,d,e}	{b,c,d,e}	{c,d,e,f}	{a,b,c,d,e,f}
{}	1	-1	-1	-1	2	-1	1	1	1	-2
{a}	0	1	0	0	-1	0	-1	0	0	1
{b}	0	0	1	0	-1	0	0	-1	0	1
{f}	0	0	0	1	-1	0	0	0	-1	1
{a,b,f}	0	0	0	0	1	0	0	0	0	-1
{c,d,e}	0	0	0	0	0	1	-1	-1	-1	2
{a,c,d,e}	0	0	0	0	0	0	1	0	0	-1
{b,c,d,e}	0	0	0	0	0	0	0	1	0	-1
{c,d,e,f}	0	0	0	0	0	0	0	0	1	-1
{a,b,c,d,e,f}	0	0	0	0	0	0	0	0	0	1

Matroid: 6-63

Basis: {a,b,c}, {a,b,d}, {a,c,d}

Rank: 3   Permutations: 60   Augmented: Y   Extended: Y

RG: $3 + 7\varphi + 5\varphi^2 + \varphi^3 + 6\theta + 13\theta\varphi + 8\theta\varphi^2 + \theta\varphi^3 + 4\theta^2 + 8\theta^2\varphi + 4\theta^2\varphi^2 + \theta^3 + 2\theta^3\varphi + \theta^3\varphi^2$

RP: $1 + 4x + 4x^2 + x^3$

$\chi: -2 + 5x - 4x^2 + x^3$

DP: $\varphi^2 + \theta^2\varphi^3 + \theta^3\varphi$

Möbius Function:

	{e,f}	{a,e,f}	{b,e,f}	{c,e,f}	{d,e,f}	{a,b,e,f}	{a,c,e,f}	{a,d,e,f}	{b,c,d,e,f}	{a,b,c,d,e,f}
{e,f}	1	-1	-1	-1	-1	1	1	1	2	-2
{a,e,f}	0	1	0	0	0	-1	-1	-1	0	2
{b,e,f}	0	0	1	0	0	-1	0	0	-1	1
{c,e,f}	0	0	0	1	0	0	-1	0	-1	1
{d,e,f}	0	0	0	0	1	0	0	-1	-1	1
{a,b,e,f}	0	0	0	0	0	1	0	0	0	-1
{a,c,e,f}	0	0	0	0	0	0	1	0	0	-1
{a,d,e,f}	0	0	0	0	0	0	0	1	0	-1
{b,c,d,e,f}	0	0	0	0	0	0	0	0	1	-1
{a,b,c,d,e,f}	0	0	0	0	0	0	0	0	0	1

Matroid: 6-64

Basis: {a,b,c}, {a,b,d}, {a,c,d}, {b,c,d}

Rank: 3   Permutations: 15   Augmented: N   Extended: Y

RG: $4 + 9\varphi + 6\varphi^2 + \varphi^3 + 6\theta + 12\theta\varphi + 6\theta\varphi^2 + 4\theta^2 + 8\theta^2\varphi + 4\theta^2\varphi^2 + \theta^3 + 2\theta^3\varphi + \theta^3\varphi^2$

RP: $1 + 6x + 4x^2 + x^3$

$\chi: -3 + 6x - 4x^2 + x^3$

DP: $\varphi^2 + \theta^3\varphi^4$

Matroid: 6-65

Basis: {a,b,c}, {a,b,d}, {a,c,d}, {b,c,e}, {b,d,e}, {c,d,e}

Rank: 3   Permutations: 60   Augmented: N   Extended: Y

RG: $6 + 11\varphi + 6\varphi^2 + \varphi^3 + 9\theta + 13\theta\varphi + 4\theta\varphi^2 + 5\theta^2 + 6\theta^2\varphi + \theta^2\varphi^2 + \theta^3 + \theta^3\varphi$

RP: $1 + 4x + 4x^2 + x^3$

$\chi: -2 + 5x - 4x^2 + x^3$

DP: $\varphi + \theta^3\varphi^2 + \theta^4\varphi^3$

Matroid: 6-66

Basis: $\{a,b,c\}$, $\{a,b,d\}$, $\{a,c,e\}$, $\{a,d,e\}$

Rank: 3   Permutations: 90   Augmented: Y   Extended: Y

RG: $4 + 8\varphi + 5\varphi^2 + \varphi^3 + 8\theta + 14\theta\varphi + 7\theta\varphi^2 + \theta\varphi^3 + 5\theta^2 + 7\theta^2\varphi + 2\theta^2\varphi^2 + \theta^3 + \theta^3\varphi$

RP: $1 + 3x + 3x^2 + x^3$

$\chi: -1 + 3x - 3x^2 + x^3$

DP: $\varphi + 2\theta^2\varphi^2 + \theta^4\varphi$

Möbius Function:

	{f}	{a,f}	{b,e,f}	{c,d,f}	{a,b,e,f}	{a,c,d,f}	{b,c,d,e,f}	{a,b,c,d,e,f}
{f}	1	-1	-1	-1	1	1	1	-1
{a,f}	0	1	0	0	-1	-1	0	1
{b,e,f}	0	0	1	0	-1	0	-1	1
{c,d,f}	0	0	0	1	0	-1	-1	1
{a,b,e,f}	0	0	0	0	1	0	0	-1
{a,c,d,f}	0	0	0	0	0	1	0	-1
{b,c,d,e,f}	0	0	0	0	0	0	1	-1
{a,b,c,d,e,f}	0	0	0	0	0	0	0	1

Matroid: 6-67

Basis: $\{a,b,c\}$, $\{a,b,d\}$, $\{a,c,e\}$, $\{a,d,e\}$, $\{b,c,f\}$, $\{b,d,f\}$, $\{c,e,f\}$, $\{d,e,f\}$

Rank: 3   Permutations: 15   Augmented: N   Extended: N

RG: $8 + 12\varphi + 6\varphi^2 + \varphi^3 + 12\theta + 12\theta\varphi + 3\theta\varphi^2 + 6\theta^2 + 3\theta^2\varphi + \theta^3$

RP: $1 + 3x + 3x^2 + x^3$

$\chi: -1 + 3x - 3x^2 + x^3$

DP: $3\theta^4\varphi^2$

Möbius Function:

	{}	{a,f}	{b,e}	{c,d}	{a,b,e,f}	{a,c,d,f}	{b,c,d,e}	{a,b,c,d,e,f}
{}	1	-1	-1	-1	1	1	1	-1
{a,f}	0	1	0	0	-1	-1	0	1
{b,e}	0	0	1	0	-1	0	-1	1
{c,d}	0	0	0	1	0	-1	-1	1
{a,b,e,f}	0	0	0	0	1	0	0	-1
{a,c,d,f}	0	0	0	0	0	1	0	-1
{b,c,d,e}	0	0	0	0	0	0	1	-1
{a,b,c,d,e,f}	0	0	0	0	0	0	0	1

Matroid: 6-68

Basis: $\{a,b,c,d\}$

Rank: 4   Permutations: 15   Augmented: Y   Extended: Y

RG: $1 + 2\varphi + \varphi^2 + 4\theta + 8\theta\varphi + 4\theta\varphi^2 + 6\theta^2 + 12\theta^2\varphi + 6\theta^2\varphi^2 + 4\theta^3 + 8\theta^3\varphi + 4\theta^3\varphi^2 + \theta^4 + 2\theta^4\varphi + \theta^4\varphi^2$

RP: $1 + 4x + 6x^2 + 4x^3 + x^4$

$\chi: 1 - 4x + 6x^2 - 4x^3 + x^4$

DP: $\varphi^2 + \theta\varphi^4$

Matroid: 6-69

Basis: {a,b,c,d}, {a,b,c,e}

Rank: 4   Permutations: 60   Augmented: Y   Extended: Y

RG: $2 + 3\varphi + \varphi^2 + 7\theta + 10\theta\varphi + 3\theta\varphi^2 + 9\theta^2 + 12\theta^2\varphi + 3\theta^2\varphi^2 + 5\theta^3 + 6\theta^3\varphi + \theta^3\varphi^2 + \theta^4 + \theta^4\varphi$

RP: $1 + 4x + 6x^2 + 4x^3 + x^4$

χ : $1 - 4x + 6x^2 - 4x^3 + x^4$

DP: $\varphi + \theta\varphi^2 + \theta^2\varphi^3$

Matroid: 6-70

Basis: {a,b,c,d}, {a,b,c,e}, {a,b,c,f}

Rank: 4   Permutations: 20   Augmented: Y   Extended: N

RG: $3 + 3\varphi + \varphi^2 + 10\theta + 9\theta\varphi + 3\theta\varphi^2 + 12\theta^2 + 9\theta^2\varphi + 3\theta^2\varphi^2 + 6\theta^3 + 3\theta^3\varphi + \theta^3\varphi^2 + \theta^4$

RP: $1 + 4x + 6x^2 + 4x^3 + x^4$

χ : $1 - 4x + 6x^2 - 4x^3 + x^4$

DP: $\theta\varphi^3 + \theta^3\varphi^3$

Matroid: 6-71

Basis: {a,b,c,d}, {a,b,c,e}, {a,b,c,f}, {a,b,d,e}, {a,b,d,f}

Rank: 4   Permutations: 90   Augmented: Y   Extended: N

RG: $5 + 4\varphi + \varphi^2 + 14\theta + 9\theta\varphi + 2\theta\varphi^2 + 14\theta^2 + 6\theta^2\varphi + \theta^2\varphi^2 + 6\theta^3 + \theta^3\varphi + \theta^4$

RP: $1 + 5x + 8x^2 + 5x^3 + x^4$

χ : $2 - 7x + 9x^2 - 5x^3 + x^4$

DP: $\theta^2\varphi^2 + \theta^3\varphi^2 + \theta^5\varphi^2$

Matroid: 6-72

Basis: {a,b,c,d}, {a,b,c,e}, {a,b,c,f}, {a,b,d,e}, {a,b,d,f}, {a,b,e,f}

Rank: 4   Permutations: 15   Augmented: Y   Extended: N

RG: $6 + 4\varphi + \varphi^2 + 16\theta + 8\theta\varphi + 2\theta\varphi^2 + 15\theta^2 + 4\theta^2\varphi + \theta^2\varphi^2 + 6\theta^3 + \theta^4$

RP: $1 + 6x + 10x^2 + 6x^3 + x^4$

χ : $3 - 10x + 12x^2 - 6x^3 + x^4$

DP: $\theta^3\varphi^4 + \theta^6\varphi^2$

Matroid: 6-73

Basis: {a,b,c,d}, {a,b,c,e}, {a,b,c,f}, {a,b,d,e}, {a,b,d,f}, {a,b,e,f}, {a,c,d,e}, {a,c,d,f}, {a,c,e,f}

Rank: 4   Permutations: 60   Augmented: Y   Extended: N

RG: $9 + 5\varphi + \varphi^2 + 19\theta + 6\theta\varphi + \theta\varphi^2 + 15\theta^2 + \theta^2\varphi + 6\theta^3 + \theta^4$

RP: $1 + 9x + 13x^2 + 6x^3 + x^4$

$\chi: 5 - 14x + 14x^2 - 6x^3 + x^4$

DP: $\theta^5\varphi^3 + \theta^6\varphi^2 + \theta^9\varphi$

Matroid: 6-74

Basis: {a,b,c,d}, {a,b,c,e}, {a,b,c,f}, {a,b,d,e}, {a,b,d,f}, {a,b,e,f}, {a,c,d,e}, {a,c,d,f}, {a,c,e,f}, {a,d,e,f}

Rank: 4   Permutations: 6   Augmented: Y   Extended: N

RG: $10 + 5\varphi + \varphi^2 + 20\theta + 5\theta\varphi + \theta\varphi^2 + 15\theta^2 + 6\theta^3 + \theta^4$

RP: $1 + 11x + 15x^2 + 6x^3 + x^4$

$\chi: 6 - 16x + 15x^2 - 6x^3 + x^4$

DP: $\theta^6\varphi^5 + \theta^{10}\varphi$

Matroid: 6-75

Basis: {a,b,c,d}, {a,b,c,e}, {a,b,c,f}, {a,b,d,e}, {a,b,d,f}, {a,b,e,f}, {a,c,d,e}, {a,c,d,f}, {a,c,e,f}, {a,d,e,f}, {b,c,d,e}, {b,c,d,f}, {b,c,e,f}, {b,d,e,f}

Rank: 4   Permutations: 15   Augmented: N   Extended: N

RG: $14 + 6\varphi + \varphi^2 + 20\theta + \theta\varphi + 15\theta^2 + 6\theta^3 + \theta^4$

RP: $1 + 17x + 15x^2 + 6x^3 + x^4$

$\chi: 9 - 19x + 15x^2 - 6x^3 + x^4$

DP: $\theta^9\varphi^4 + \theta^{10}\varphi^2$

Matroid: 6-76

Basis: {a,b,c,d}, {a,b,c,e}, {a,b,c,f}, {a,b,d,e}, {a,b,d,f}, {a,b,e,f}, {a,c,d,e}, {a,c,d,f}, {a,c,e,f}, {a,d,e,f}, {b,c,d,e}, {b,c,d,f}, {b,c,e,f}, {b,d,e,f}, {c,d,e,f}

Rank: 4   Permutations: 1   Augmented: N   Extended: N

RG: $15 + 6\varphi + \varphi^2 + 20\theta + 15\theta^2 + 6\theta^3 + \theta^4$

RP: $1 + 20x + 15x^2 + 6x^3 + x^4$

$\chi: 10 - 20x + 15x^2 - 6x^3 + x^4$

DP: $\theta^{10}\varphi^6$

Matroid: 6-77

Basis: {a,b,c,d}, {a,b,c,e}, {a,b,c,f}, {a,b,d,e}, {a,b,d,f}, {a,b,e,f}, {a,c,d,e}, {a,c,d,f}, {a,c,e,f}, {b,c,d,e}, {b,c,d,f}, {b,c,e,f}

Rank: 4   Permutations: 20   Augmented: N   Extended: N

RG: $12 + 6\varphi + \varphi^2 + 19\theta + 3\theta\varphi + 15\theta^2 + \theta^2\varphi + 6\theta^3 + \theta^4$

RP: $1 + 13x + 13x^2 + 6x^3 + x^4$

$\chi: 7 - 16x + 14x^2 - 6x^3 + x^4$

DP: $\theta^7\varphi^3 + \theta^9\varphi^3$

Matroid: 6-78

Basis: {a,b,c,d}, {a,b,c,e}, {a,b,c,f}, {a,b,d,e}, {a,b,d,f}, {a,b,e,f}, {a,c,d,e}, {a,c,d,f}, {a,c,e,f}, {b,c,d,e}, {b,c,d,f}, {b,d,e,f}, {c,d,e,f}

Rank: 4   Permutations: 45   Augmented: N   Extended: N

RG: $13 + 6\varphi + \varphi^2 + 20\theta + 2\theta\varphi + 15\theta^2 + 6\theta^3 + \theta^4$

RP: $1 + 14x + 15x^2 + 6x^3 + x^4$

$\chi: 8 - 18x + 15x^2 - 6x^3 + x^4$

DP: $\theta^8\varphi^2 + 2\theta^9\varphi^2$

Matroid: 6-79

Basis: {a,b,c,d}, {a,b,c,e}, {a,b,c,f}, {a,b,d,e}, {a,b,d,f}, {a,c,d,e}, {a,c,d,f}

Rank: 4   Permutations: 60   Augmented: Y   Extended: N

RG: $7 + 5\varphi + \varphi^2 + 16\theta + 8\theta\varphi + \theta\varphi^2 + 14\theta^2 + 4\theta^2\varphi + 6\theta^3 + \theta^3\varphi + \theta^4$

RP: $1 + 7x + 10x^2 + 5x^3 + x^4$

$\chi: 3 - 9x + 10x^2 - 5x^3 + x^4$

DP: $\theta^3\varphi^2 + \theta^5\varphi^3 + \theta^7\varphi$

Matroid: 6-80

Basis: {a,b,c,d}, {a,b,c,e}, {a,b,c,f}, {a,b,d,e}, {a,b,d,f}, {a,c,d,e}, {a,c,d,f}, {b,c,d,e}, {b,c,d,f}

Rank: 4   Permutations: 15   Augmented: N   Extended: N

RG: $9 + 6\varphi + \varphi^2 + 16\theta + 6\theta\varphi + 14\theta^2 + 4\theta^2\varphi + 6\theta^3 + \theta^3\varphi + \theta^4$

RP: $1 + 10x + 10x^2 + 5x^3 + x^4$

$\chi: 4 - 10x + 10x^2 - 5x^3 + x^4$

DP: $\theta^4\varphi^2 + \theta^7\varphi^4$

Matroid: 6-81

Basis: {a,b,c,d}, {a,b,c,e}, {a,b,c,f}, {a,b,d,e}, {a,b,d,f}, {a,c,d,e}, {a,c,d,f}, {b,c,d,e}, {b,c,e,f}, {b,d,e,f}, {c,d,e,f}

Rank: 4   Permutations: 60   Augmented: N   Extended: N

RG: $11 + 6\varphi + \varphi^2 + 19\theta + 4\theta\varphi + 15\theta^2 + \theta^2\varphi + 6\theta^3 + \theta^4$

RP: $1 + 10x + 13x^2 + 6x^3 + x^4$

$\chi : 6 - 15x + 14x^2 - 6x^3 + x^4$

DP: $\theta^6\varphi + \theta^7\varphi^2 + \theta^8\varphi^3$

Matroid: 6-82

Basis: {a,b,c,d}, {a,b,c,e}, {a,b,c,f}, {a,b,d,e}, {a,b,d,f}, {a,c,d,e}, {a,c,e,f}, {a,d,e,f}

Rank: 4   Permutations: 90   Augmented: Y   Extended: N

RG: $8 + 5\varphi + \varphi^2 + 18\theta + 7\theta\varphi + \theta\varphi^2 + 15\theta^2 + 2\theta^2\varphi + 6\theta^3 + \theta^4$

RP: $1 + 7x + 11x^2 + 6x^3 + x^4$

$\chi : 4 - 12x + 13x^2 - 6x^3 + x^4$

DP: $\theta^4\varphi + 2\theta^5\varphi^2 + \theta^8\varphi$

Matroid: 6-83

Basis: {a,b,c,d}, {a,b,c,e}, {a,b,c,f}, {a,b,d,e}, {a,b,d,f}, {a,c,d,e}, {a,c,e,f}, {a,d,e,f}, {b,c,d,f}, {b,c,e,f}, {b,d,e,f}, {c,d,e,f}

Rank: 4   Permutations: 15   Augmented: N   Extended: N

RG: $12 + 6\varphi + \varphi^2 + 20\theta + 3\theta\varphi + 15\theta^2 + 6\theta^3 + \theta^4$

RP: $1 + 11x + 15x^2 + 6x^3 + x^4$

$\chi : 7 - 17x + 15x^2 - 6x^3 + x^4$

DP: $3\theta^8\varphi^2$

Matroid: 6-84

Basis: {a,b,c,d}, {a,b,c,e}, {a,b,d,e}

Rank: 4   Permutations: 60   Augmented: Y   Extended: Y

RG: $3 + 4\varphi + \varphi^2 + 9\theta + 11\theta\varphi + 2\theta\varphi^2 + 10\theta^2 + 11\theta^2\varphi + \theta^2\varphi^2 + 5\theta^3 + 5\theta^3\varphi + \theta^4 + \theta^4\varphi$

RP: $1 + 5x + 8x^2 + 5x^3 + x^4$

$\chi : 2 - 7x + 9x^2 - 5x^3 + x^4$

DP: $\varphi + \theta^2\varphi^3 + \theta^3\varphi^2$

Matroid: 6-85
Basis: {a,b,c,d}, {a,b,c,e}, {a,b,d,e}, {a,c,d,e}
Rank: 4   Permutations: 30   Augmented: Y   Extended: Y
RG: $4 + 5\varphi + \varphi^2 + 10\theta + 11\theta\varphi + \theta\varphi^2 + 10\theta^2 + 10\theta^2\varphi + 5\theta^3 + 5\theta^3\varphi + \theta^4 + \theta^4\varphi$
RP: $1 + 7x + 10x^2 + 5x^3 + x^4$
$\chi: 3 - 9x + 10x^2 - 5x^3 + x^4$
DP: $\varphi + \theta^3\varphi^4 + \theta^4\varphi$

Matroid: 6-86
Basis: {a,b,c,d}, {a,b,c,e}, {a,b,d,e}, {a,c,d,e}, {b,c,d,e}
Rank: 4   Permutations: 6   Augmented: N   Extended: Y
RG: $5 + 6\varphi + \varphi^2 + 10\theta + 10\theta\varphi + 10\theta^2 + 10\theta^2\varphi + 5\theta^3 + 5\theta^3\varphi + \theta^4 + \theta^4\varphi$
RP: $1 + 10x + 10x^2 + 5x^3 + x^4$
$\chi: 4 - 10x + 10x^2 - 5x^3 + x^4$
DP: $\varphi + \theta^4\varphi^5$

Matroid: 6-87
Basis: {a,b,c,d}, {a,b,c,e}, {a,b,d,e}, {a,c,d,e}, {b,c,d,f}, {b,c,e,f}, {b,d,e,f}, {c,d,e,f}
Rank: 4   Permutations: 15   Augmented: N   Extended: N
RG: $8 + 6\varphi + \varphi^2 + 16\theta + 7\theta\varphi + 14\theta^2 + 4\theta^2\varphi + 6\theta^3 + \theta^3\varphi + \theta^4$
RP: $1 + 7x + 10x^2 + 5x^3 + x^4$
$\chi: 3 - 9x + 10x^2 - 5x^3 + x^4$
DP: $\theta^4\varphi^2 + \theta^6\varphi^4$

Matroid: 6-88
Basis: {a,b,c,d}, {a,b,c,e}, {a,b,d,e}, {a,c,d,f}, {a,c,e,f}, {a,d,e,f}
Rank: 4   Permutations: 60   Augmented: Y   Extended: N
RG: $6 + 5\varphi + \varphi^2 + 15\theta + 9\theta\varphi + \theta\varphi^2 + 14\theta^2 + 5\theta^2\varphi + 6\theta^3 + \theta^3\varphi + \theta^4$
RP: $1 + 5x + 8x^2 + 5x^3 + x^4$
$\chi: 2 - 7x + 9x^2 - 5x^3 + x^4$
DP: $\theta^3\varphi^2 + \theta^4\varphi^3 + \theta^6\varphi$

Matroid: 6-89

Basis: {a,b,c,d}, {a,b,c,e}, {a,b,d,e}, {a,c,d,f}, {a,c,e,f}, {a,d,e,f}, {b,c,d,f}, {b,c,e,f}, {b,d,e,f}

Rank: 4   Permutations: 10   Augmented: N   Extended: N

RG: $9 + 6\varphi + \varphi^2 + 18\theta + 6\theta\varphi + 15\theta^2 + 2\theta^2\varphi + 6\theta^3 + \theta^4$

RP: $1 + 6x + 11x^2 + 6x^3 + x^4$

$\chi : 4 - 12x + 13x^2 - 6x^3 + x^4$

DP: $2\theta^6\varphi^3$

Matroid: 6-90

Basis: {a,b,c,d}, {a,b,c,e}, {a,b,d,f}, {a,b,e,f}

Rank: 4   Permutations: 45   Augmented: Y   Extended: N

RG: $4 + 4\varphi + \varphi^2 + 12\theta + 10\theta\varphi + 2\theta\varphi^2 + 13\theta^2 + 8\theta^2\varphi + \theta^2\varphi^2 + 6\theta^3 + 2\theta^3\varphi + \theta^4$

RP: $1 + 4x + 6x^2 + 4x^3 + x^4$

$\chi : 1 - 4x + 6x^2 - 4x^3 + x^4$

DP: $2\theta^2\varphi^2 + \theta^4\varphi^2$

Matroid: 6-91

Basis:

Rank: 0   Permutations: 1   Augmented: N   Extended: Y

RG: $1 + 6\varphi + 15\varphi^2 + 20\varphi^3 + 15\varphi^4 + 6\varphi^5 + \varphi^6$

RP: 1

$\chi : 1$

DP: φ^6

Möbius Function:

                {a,b,c,d,e,f}

{a,b,c,d,e,f}   1

Matroid: 6-92

Basis: {a,b,c,d,e}

Rank: 5   Permutations: 6   Augmented: Y   Extended: Y

RG: $1 + \varphi + 5\theta + 5\theta\varphi + 10\theta^2 + 10\theta^2\varphi + 10\theta^3 + 10\theta^3\varphi + 5\theta^4 + 5\theta^4\varphi + \theta^5 + \theta^5\varphi$

RP: $1 + 5x + 10x^2 + 10x^3 + 5x^4 + x^5$

$\chi : -1 + 5x - 10x^2 + 10x^3 - 5x^4 + x^5$

DP: $\varphi + \theta\varphi^5$

Matroid: 6-93

Basis: {a,b,c,d,e}, {a,b,c,d,f}

Rank: 5   Permutations: 15   Augmented: Y   Extended: N

RG: $2 + \varphi + 9\theta + 4\theta\varphi + 16\theta^2 + 6\theta^2\varphi + 14\theta^3 + 4\theta^3\varphi + 6\theta^4 + \theta^4\varphi + \theta^5$

RP: $1 + 5x + 10x^2 + 10x^3 + 5x^4 + x^5$

$\chi : -1 + 5x - 10x^2 + 10x^3 - 5x^4 + x^5$

DP: $\theta\varphi^2 + \theta^2\varphi^4$

Matroid: 6-94

Basis: {a,b,c,d,e}, {a,b,c,d,f}, {a,b,c,e,f}

Rank: 5   Permutations: 20   Augmented: Y   Extended: N

RG: $3 + \varphi + 12\theta + 3\theta\varphi + 19\theta^2 + 3\theta^2\varphi + 15\theta^3 + \theta^3\varphi + 6\theta^4 + \theta^5$

RP: $1 + 6x + 13x^2 + 13x^3 + 6x^4 + x^5$

$\chi : -2 + 9x - 16x^2 + 14x^3 - 6x^4 + x^5$

DP: $\theta^2\varphi^3 + \theta^3\varphi^3$

Matroid: 6-95

Basis: {a,b,c,d,e}, {a,b,c,d,f}, {a,b,c,e,f}, {a,b,d,e,f}

Rank: 5   Permutations: 15   Augmented: Y   Extended: N

RG: $4 + \varphi + 14\theta + 2\theta\varphi + 20\theta^2 + \theta^2\varphi + 15\theta^3 + 6\theta^4 + \theta^5$

RP: $1 + 8x + 17x^2 + 15x^3 + 6x^4 + x^5$

$\chi : -3 + 12x - 19x^2 + 15x^3 - 6x^4 + x^5$

DP: $\theta^3\varphi^4 + \theta^4\varphi^2$

Matroid: 6-96

Basis: {a,b,c,d,e}, {a,b,c,d,f}, {a,b,c,e,f}, {a,b,d,e,f}, {a,c,d,e,f}

Rank: 5   Permutations: 6   Augmented: Y   Extended: N

RG: $5 + \varphi + 15\theta + \theta\varphi + 20\theta^2 + 15\theta^3 + 6\theta^4 + \theta^5$

RP: $1 + 11x + 20x^2 + 15x^3 + 6x^4 + x^5$

$\chi : -4 + 14x - 20x^2 + 15x^3 - 6x^4 + x^5$

DP: $\theta^4\varphi^5 + \theta^5\varphi$

Matroid: 6-97

Basis: {a,b,c,d,e}, {a,b,c,d,f}, {a,b,c,e,f}, {a,b,d,e,f}, {a,c,d,e,f}, {b,c,d,e,f}

Rank: 5   Permutations: 1   Augmented: N   Extended: N

RG: $6 + \varphi + 15\theta + 20\theta^2 + 15\theta^3 + 6\theta^4 + \theta^5$

RP: $1 + 15x + 20x^2 + 15x^3 + 6x^4 + x^5$

χ : $-5 + 15x - 20x^2 + 15x^3 - 6x^4 + x^5$

DP: $\theta^5\varphi^6$

Matroid: 6-98

Basis: {a,b,c,d,e,f}

Rank: 6   Permutations: 1   Augmented: Y   Extended: N

RG: $1 + 6\theta + 15\theta^2 + 20\theta^3 + 15\theta^4 + 6\theta^5 + \theta^6$

RP: $1 + 6x + 15x^2 + 20x^3 + 15x^4 + 6x^5 + x^6$

χ : $1 - 6x + 15x^2 - 20x^3 + 15x^4 - 6x^5 + x^6$

DP: $\theta\varphi^6$

Index

www.ingramcontent.com/pod-product-compliance
Lightning Source LLC
Chambersburg PA
CBHW060324200326
41519CB00011BA/1827